Li Niu, Dongxue Han

Chemical Analysis of Antioxidant Capacity

Also of Interest

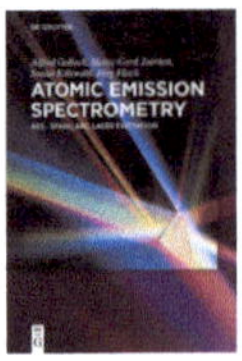

Atomic Emission Spectrometry.
AES – Spark, Arc, Laser Excitation
Golloch, Joosten, Killewald, Flock, 2019
ISBN 978-3-11-052768-1, e-ISBN 978-3-11-052969-2

Elemental Analysis.
An Introduction to Modern Spectrometric Techniques
Schlemmer, Balcaen, Todolí, Hinds, 2018
ISBN 978-3-11-050107-0, e-ISBN 978-3-11-050108-7

Organic Trace Analysis.
Niessner, Schäffer, 2017
ISBN 978-3-11-044114-7, e-ISBN 978-3-11-044115-4

Inorganic Trace Analytics.
Trace Element Analysis and Speciation
Matusiewicz, Bulska (Eds.), 2017
ISBN 978-3-11-037194-9, e-ISBN 978-3-11-036673-0

Li Niu, Dongxue Han

Chemical Analysis of Antioxidant Capacity

Mechanisms and Techniques

DE GRUYTER

科学出版社

Authors
Prof. Li Niu
Center for Advanced Analytical Science
School of Chemistry and Chemical Engineering
Guangzhou University
Guangzhou 510006
P.R. China
lniu@gzhu.edu.cn

Prof. Dongxue Han
Center for Advanced Analytical Science
School of Chemistry and Chemical Engineering
Guangzhou University
Guangzhou 510006
P.R. China
dxhan@gzhu.edu.cn

ISBN 978-3-11-057314-5
e-ISBN (PDF) 978-3-11-057376-3
e-ISBN (EPUB) 978-3-11-057323-7

Library of Congress Control Number: 2018964148

Bibliographic information published by the Deutsche Nationalbibliothek
The Deutsche Nationalbibliothek lists this publication in the Deutsche Nationalbibliografie; detailed bibliographic data are available on the Internet at http://dnb.dnb.de.

Typesetting: Integra Software Services Pvt. Ltd.
Printing and binding: CPI books GmbH, Leck

www.degruyter.com

Contents

1 Oxidation and antioxidation

Oxygen is essential for life. Although oxygen is indispensable for aerobic organisms for their growth and reproduction, it can also be considered as a "dangerous friend" ascribed to its free radicals trigger as well as oxidative stress during the process of aerobic respiration, which seriously affects human health and life expectancy.

Free radicals refer to atoms or molecules that contain an odd number of free unpaired electron(s). A large number of free radicals are obtained from the metabolism of organisms, which usually have a very short lifetime, appear lively and unstable, and exhibit paramagnetic features. They result in a chain reaction to generate new free radicals or oxides. To maintain a normal operation of the body, there are some in vivo water- or fat-soluble antioxidants that can effectively eliminate free radicals to ensure a balance of oxidative stress and oxidative defense (Figure 1.1). When a body's antioxidant mechanism is not functioning well, the excess of free radicals generated can damage a number of biomolecules including lipids, proteins, carbohydrates as well as nucleic acids, leading to a decline in health and various diseases.

1.1 Biological activity of oxidation in the human body

In vivo biologically active substances usually have strong oxidative nature, which is a result of aerobic cellular respiration, metabolism, and mitochondrial electron transfer chain process. Typically, they can be classified as reactive oxygen species (ROS) and reactive nitrogen species (RNS). ROS includes oxygen-containing neutral molecules (H_2O_2, singlet oxygen, and hypohalous acid), oxygen-containing free radicals (hydroxyl, hydrogen peroxide, alkyl peroxy, and aliphatic alkoxy radicals), and oxygen-containing ions (superoxide anion). In the process of biological evolution, oxygen is the terminal electron acceptor during respiration. Although oxygen has a double free radical, its two unpaired electrons are in the same spin direction; hence, oxygen is not dangerous. Among the oxygen neutral molecules, H_2O_2 is a cell membrane-permeable and relatively stable, active substance, which is generally produced by peroxisome. In general, H_2O_2 does not directly damage the lipid or DNA, but via the Fenton reaction it can react with metal ions (such as copper or ferrous ions) to form a complex and thus generates hydroxyl radicals with high activity. In addition, peroxidase in neutrophils can catalyze the reaction of H_2O_2, with physiological concentrations of chloride, to produce hypochlorous acid. About 2% of oxygen inhaled during the respiration, which is the basic process of a life, gets converted to superoxide anion [1]. The half-life of superoxide is longer than that of other free radicals, but its biological activity is far less than that of hydroxyl free radicals. When the superoxide anion disproportionation reaction takes place in water, electronically excited

https://doi.org/10.1515/9783110573763-001

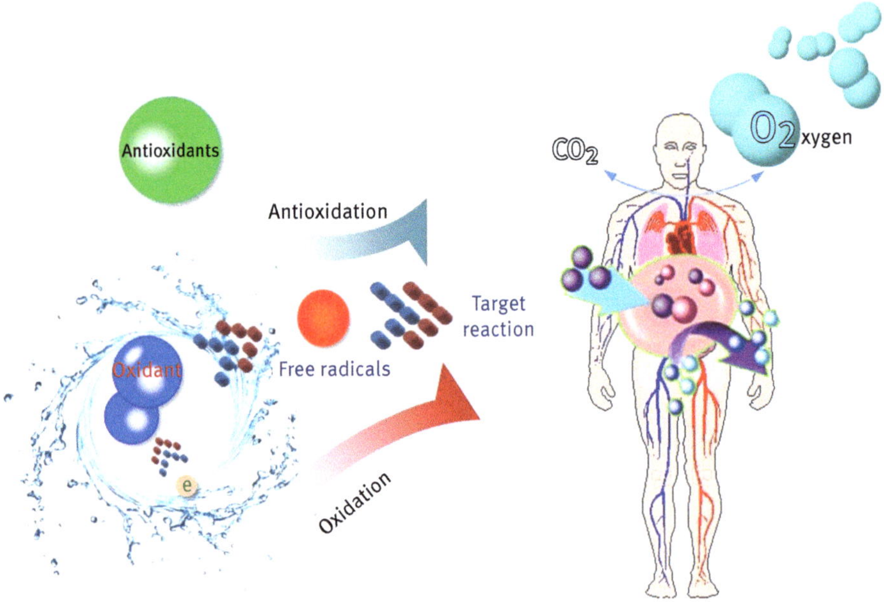

Figure 1.1: Oxidation and antioxidation in a human body.

singlet oxygen can be obtained. In fact, the hydroxyl radicals or superoxide anions can interact with the surrounding biomolecules and can undergo a chain reaction to result in hydrogen peroxide, alkyl peroxy, and aliphatic alkoxy radicals; all these play significant roles in oxidative stress induction [2]. The process of generation and conversion of ROS is shown in Figure 1.2.

RNS mainly exists in the form of nitric oxide (NO) and a series of nitrogenous and nitro compounds (such as ONOO· and HOONO) with high oxidation activity. NO is produced by L-arginine using various nitric oxide synthase (NOS). NO can not only react with some molecules to obtain related reactive nitrogen but can also bind with a transition metal, such as iron ions, and thus plays an important role of forming guanosinic acid in the second messenger ring [3]. Mammalian cells contain NOS (NOS1, NOS2, and NOS3) encoded by three genes, which accounts to 51%–57% of the gene homology between the three isomers. NOS1 (also called nNOS, first purified from nerve cells) and NOS3 (also known as eNOS, purified from endothelial cells for the first time) from the nerve and endothelial cells carry out gene expression; it is considered as the basic form of NOS.

With an increase in the levels of tissue calcium, NOS1 and NOS3 get activated, producing transient low concentrations of NO. In contrast, NOS2 is an inducible NOS (also known as iNOS), which is calcium independent [4]. Inflammation occurs during iNOS-induced synthesis and leads to generation of excessive NO, which cause greater damage to the organs. The reaction of excess NO with superoxide

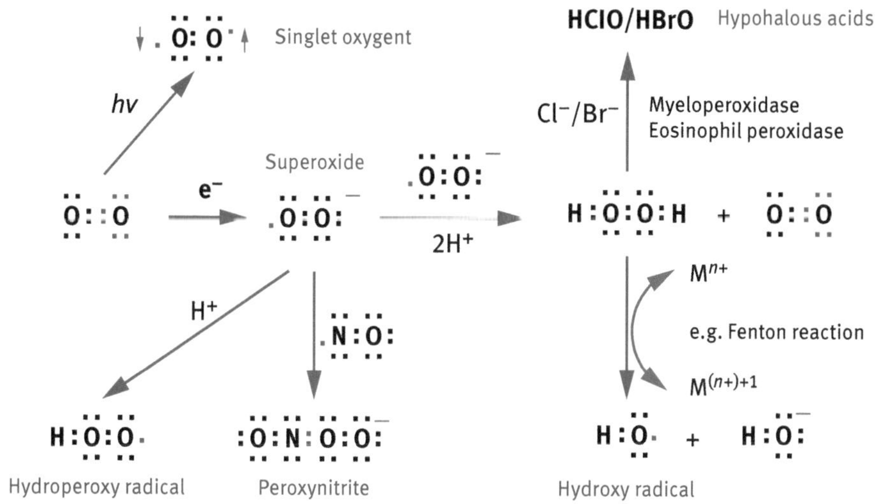

Figure 1.2: Generation and conversion process of ROS.

results in highly toxic peroxynitrite that oxidizes lipids, proteins, and DNA, exacerbating the possibility of damaging the body.

During the formation of active oxidative material substance, radical generation process contains two different forms of endogenous and exogenous oxidation.

Endogenous oxidation of an active substance (Figure 1.3) involves the following: ① mitochondrial respiratory chain process, which is the main source of free radicals; ② automatic biomolecules, such as hemoglobin, myoglobin, and catecholamines, in the body can produce superoxide; ③ enzymatic reactions, similar to xanthine oxidase, lipoxygenase, and aldehyde oxidase involved in the reaction, are a major source of free radicals; ④ owing to sudden breathing at this time, phagocytic cells, such as leukocytes, consume a large amount of oxygen during phagocytosis and by the immune defense produce superoxide and H_2O_2; ⑤ body trace of metal ions and H_2O_2 reaction; ⑥ strenuous exercise can activate xanthine oxidase, resulting in excessive free radicals and increased oxidative stress; ⑦ infection –to resist the invasion of microorganisms the immune system will generate explosive free radicals; ⑧ ischemia or reperfusion – it can also activate xanthine oxidase and can lead to the formation of oxidative active substances.

In contrast, exogenous oxidative active substances (Figure 1.4) are a result the following: ① air pollution, including primary pollution and secondary pollution. Primary pollution refers to the first generation of air pollutants, whereas secondary pollution refers to a pollutant in the atmosphere after the reaction of the resulting products. Some of the common pollutants are O_3, nitrogen oxides (such as N_2O and NO), SO_2, CO, volatile organic compounds (such as methane, benzene, xylene, and chlorofluorocarbons), NH_3, particulate matter, and heavy metals (Pb, Cd, Cu, Sb, Hg, Cr, Co, Se,

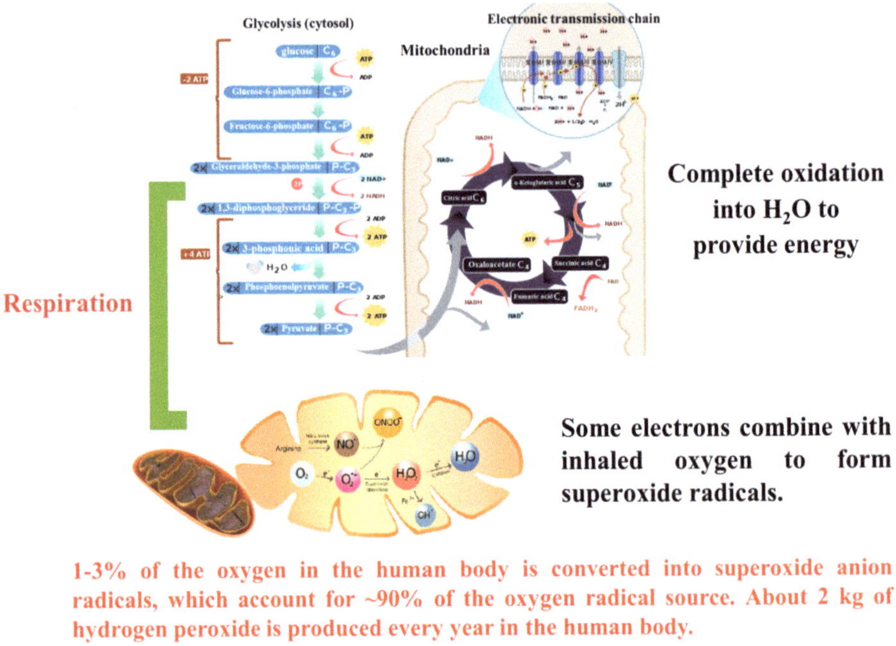

Figure 1.3: Respiration and generation of free oxygen radicals in a human body.

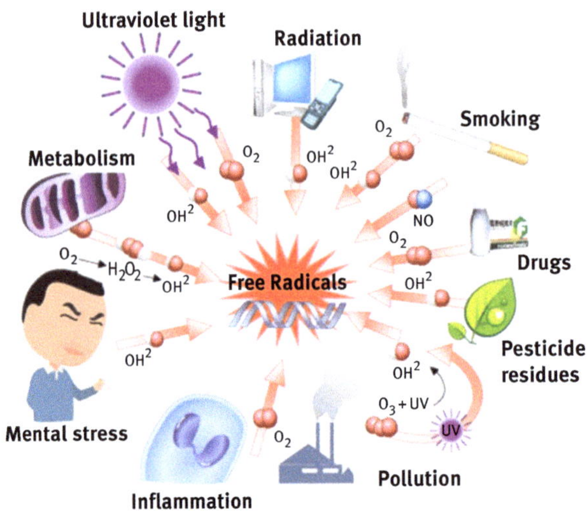

Figure 1.4: Formation of exogenous reactive oxygen–nitrogen radicals.

and Sr). A body's oxidation balance is disturbed when a person is exposed to polluted air, causing various diseases. ② Inorganic particles in the air, especially the fine dust containing minerals (such as quartz, SiO_2, and asbestos fibers), can cause oxidative stress damage. ③ Smoking is yet another factor. Tobacco smoke contains a large number of free radicals and oxidants, which when inhaled not only have a negative impact on the key biomolecules in tissues and organs but can also activate inflammatory cells, leading to additional oxidative stress. ④ Certain drugs significantly affect the body's oxidative activity: for example, analgesics (e.g., aspirin and phenacetin), anticancer drugs (e.g., methotrexate, bleomycin, doxorubicin, and topoisomerase inhibitors), immunosuppressive agents (e.g., cyclosporin A, mycophenolic acid, sirolimus, and tacrolimus), antimalarial (e.g., chloroquine, mefloquine, primaquine, and fluticasone), antibiotics (e.g., chlorine Diclofenac, Ciprofloxacin, Moxifloxacin, Nalidixic Acid, Norfloxacin, Ofloxacin, Sulfamethoxazole, Cotrimoxazole, Sulfadiazine, Sulfonamides, Sulfasalazine, and Sulfasalazine tablets), antiretroviral drugs (e.g., indinavir, atazanavir, and zidovudine), diuretics (e.g., spironolactone), narcotics (e.g., cocaine and amphetamines) and ethanol. ⑤ Inhalation of industrial solvents, such as chloroform and carbon tetrachloride, may activate cytochrome P450 that is responsible for hepatic metabolism [5], causing an imbalance of active oxidized substances. ⑥ Being in the Sun for long hours; ultraviolet radiations can cause skin oxidative stress, resulting in body dysfunction.

1.2 Oxidation and antioxidant balance

The existence of oxidized active substances is a double-edged sword; it is essential to the body but also has toxic effects (Figure 1.5). Under normal circumstances, for the formation and removal of oxidative active substances in a dynamic equilibrium, an appropriate concentration of ROS and RNS is important for normal functioning of the body.

Oxidative active substances have the following properties: ① help maintain normal oxygen respiratory metabolism, keeping the balance of redox cells; ② have a positive effect on cell signal transduction. Organisms present in the environment that change the body's functional coordination requires a unified cell communication mechanism to achieve communication between cells or to identify the presence of a variety of signals in the surrounding environment as well as the conversion by the intracellular signal transduction system, thereby changing some of the metabolic processes within the cell, affecting the growth rate of cells, and even inducing cell death. Studies have shown that most types of cells suffer from cytokines, growth factors, various interleukins, tumor necrosis factor alpha, angiotensin II, platelet growth factor, nerve growth factor, serum transforming growth factor beta 1, and granulocyte-giant. A small oxidative burst occurs when stimulated by fibroblast growth factor (FGF), resulting in low concentrations of ROS. Therefore,

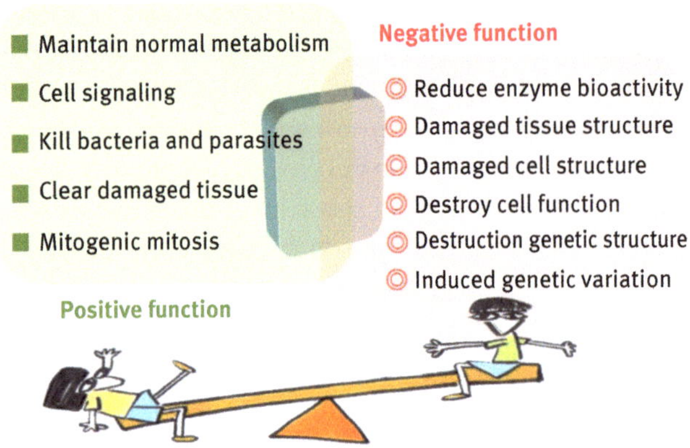

Figure 1.5: Dual nature of free radicals ROS/RNS in the body.

ROS as a second messenger plays a role in the initiation of signal transduction [6]. ③ This helps adjust the body's immune function. When inflammation or pathogen invades, immune cells turn on defenses and explosively produce a large amount of ROS to kill bacteria and parasites. ④ It helps remove damaged tissue. Autophagy is a process that engulfs its own cytoplasmic proteins or organelles and encapsulates them into vesicles, fusing with lysosomes to form autophagy lysosomes and degrading the contents they encapsulate, thereby enabling the cell's own metabolic needs. Some of these organelles are updated. As a signaling molecule inducing autophagy, ROS participates in a number of signaling pathways that induce autophagy and plays an important role in the formation of autophagy [7]. ⑤ It helps trigger cell mitosis process. Figure 1.6 shows three typical ROS generation channels.

Oxidative active substances in the positive role of the liver on the detoxification of foreign poisoning (Figure 1.7), liver microsomal cytochrome P450 catalyzed hydroxylation of various toxic substances, connected to the cytochrome O_2^- free radical is real hydroxylation of substances. The liver detoxification function is as follows. During the body's metabolic process, the portal vein collected from the abdominal cavity of blood, which is mainly the harmful substances and microbial antigenic material present in the blood, will be detoxified and cleared in the liver. The liver is the body's main detoxification organ; it can protect the body from damage such that toxic substances become nontoxic or highly soluble substances and can pass out from the body with excreted bile or urine. There are four main ways for liver detoxification: ① first one is the chemical methods, such as oxidation, reduction, decomposition, binding, and deoxygenation. Ammonia is a toxic metabolite; its detoxification is mainly by the synthesis of urea in the liver, with the excreted urine. Toxic substances as well as glucuronic acid, sulfuric acid, amino acids, and so on can be converted into nontoxic substances. ② Some heavy metals, such as mercury, as well as bacteria from the gut can be excreted while bile secretion. ③ Some alkaloids, such as

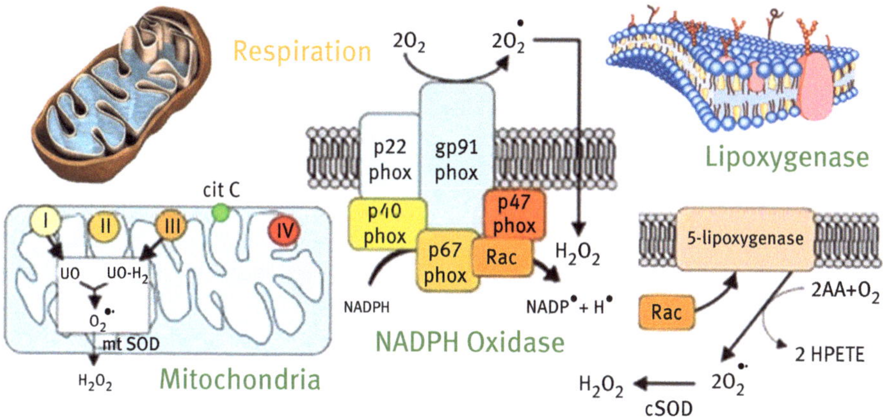

Figure 1.6: Typical ROS generation and their positive functions.

strychnine and morphine, get accumulated in the liver, thereby gradually releasing a small amount of these substances in the liver and in turn reducing the poisoning process. ④ If the liver is damaged, the body is prone to get easily poisoned or catch infection; the liver cells contain a large number of liver macrophages, which have a strong phagocytic capacity. Thus, these phagocytic bacteria protect the liver.

When the concentration of oxidized active substances is high, these are hazardous to the body, thus reducing the biological activity of enzymes, damaging tissue or cell structure, disrupting gene structure and cell function, inducing genetic variation, and so on. At this point, the body's internal antioxidant defense system is fully operational, thus minimizing the rate of oxidation of active substances and the extent of the organization's oxidative damage. The defense system includes different levels of functionality: preventive, blocking, and healing levels. The preventive level refers to the direct removal of free radicals, thus preventing them from generating free radical reactions; the blocking level refers to neutralization, thereby blocking the formation of ROS chain reaction; and the repair level refers to the repair of the damaged protein and DNA. Based on the structure and the classification mechanism of antioxidants, human antioxidant system can be divided as the primary antioxidant defense system and secondary antioxidant defense system. The primary antioxidant defense system is mainly for the free-type oxidation of active substances so as to remove or neutralize and eliminate the toxic effects on the cells. This system further classified as the enzyme antioxidant system and the nonenzyme antioxidant system. Here, the enzyme antioxidant system is mainly composed of superoxide dismutase (SOD), catalase (CAT), glutathione peroxidase (GPx), peroxide reductase (POD), and other components, and this is the body's first anti-oxidant defense. The nonenzymatic antioxidant systems rely on lipid-soluble antioxidants

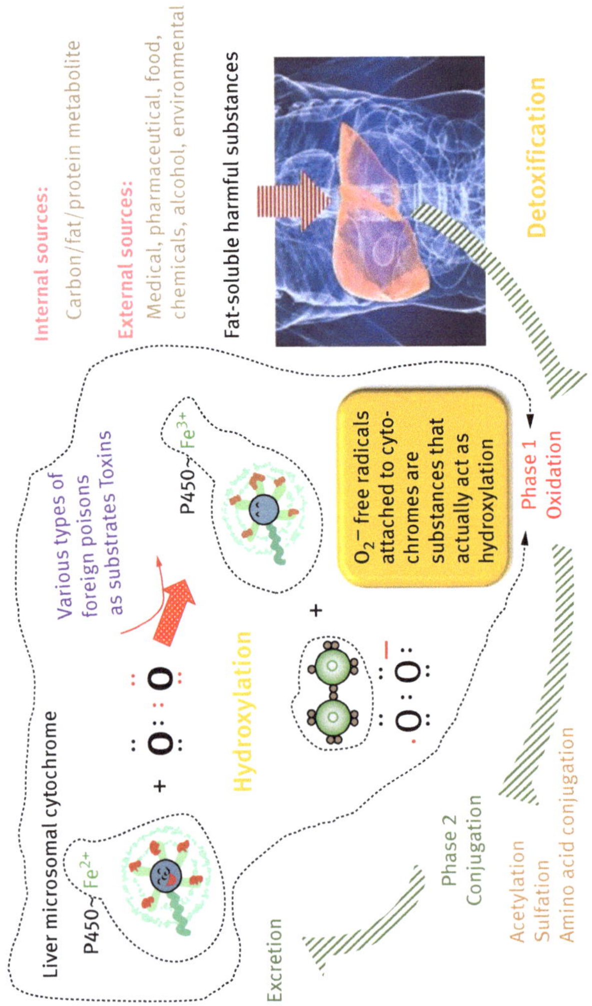

Figure 1.7: Free radical action in the process of liver detoxification.

(such as vitamin E), carotenoids and coenzymes, water-soluble antioxidants (such as vitamin C and glutathione), protein-based antioxidants (such as ferritin, Transferrin, lactoferrin, ceruloplasmin, and metallothionein), low molecular weight compounds (such as urate), melatonin, phytochemicals (such as phenols, terpenes, plant fibers, natural pigments and some traditional Chinese medicines composition), and so on. These build the body's second line of defense; most of which are obtained by in vitro intake. Secondary antioxidation royal system, also known as the antioxidant repair system, is mainly used to repair oxidatively damaged proteins and DNA, while including the body's detoxification system, membrane repair, and regeneration system. There are two ways to repair proteins: repairing or degrading damaged molecules. Studies have shown that direct repair of protein thiol groups can be achieved in systems containing glutathione or thioredoxin. In addition, methionine sulfoxide is directly repaired and reduced to methionine in the presence of methionine sulfur oxide reductase or methionine reductase. However, the repair of proteins is limited and mainly relies on the catabolism of oxidatively modified proteins by hydrolases and proteasomes. Oxidized lipids are also oxidatively metabolized into non-toxic molecules. In contrast, repair mechanisms for DNA damage are robust and easy to fix, and involve mismatch repair, direct repair, resection repair, recombination repair, and emergency response (SOS response). Figure 1.8 shows an example of

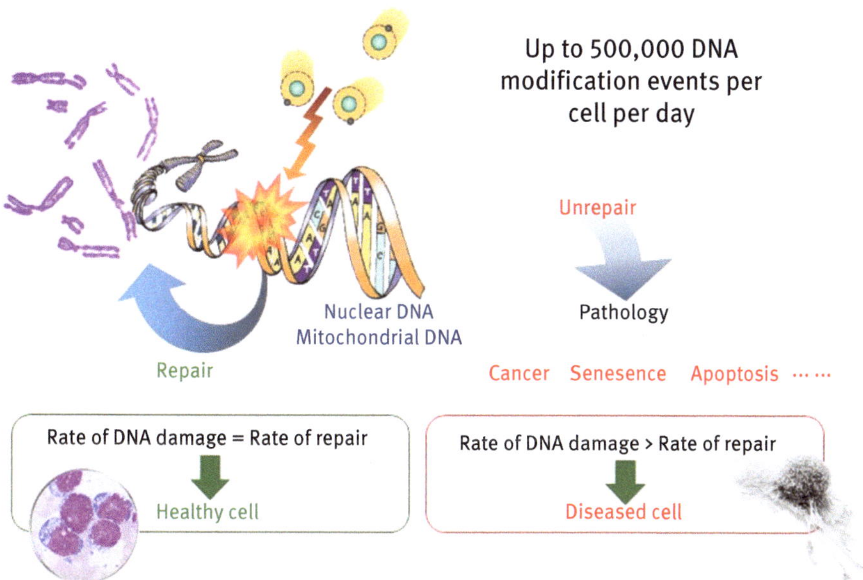

Figure 1.8: Oxidative stress leads to DNA damage.

oxidative stress leading to DNA damage. The body's antioxidant defense system is an organic whole, with a level of defense collaboration, interdependence, and complementarity. The effectiveness of the secondary antioxidant defense system depends on the normal functioning of the immune system and the primary antioxidant defense system. In the primary antioxidant defense system, the trace element selenium has GPx cofactor, closely related to its structure and function; copper, manganese, and zinc are necessary components of SOD. Therefore, the material support and supply of nonenzymatic antioxidant system laid a favorable foundation for the normal function of an enzyme antioxidant system. From this perspective, a moderate dietary supplement of body fluids in the concentration of antioxidant substances, to improve the quality of life, for antiaging is of great significance.

1.3 Diseases due to oxidation of active substances

Although the development of industrial civilization has enriched our physical and spiritual life, the accompanying by-products, namely, environmental pollution (water, air pollution, ozone depletion, ionizing radiation, etc.), unhealthy habits (excessive strenuous exercise, smoking, alcoholism, mental stress, consumption of junk food, etc.),and pathological factors (inflammation, bacterial infections, trauma, drug abuse, etc.), result in the rapid accumulation of oxidative active substances. At the same time, with an increase in age, the free radical scavenging mechanism in vivo shows a degenerative trend and cannot play a role in a timely and effective manner. Under this double negative effect, the oxidized active substances exceed the standard, and the balance between human oxidation and antioxidation is lost. A large number oxidative active substances are generated throughout the body, free from biological macromolecules. These further trigger a chain reaction, and this state is also called oxidative stress. The disease may follow when the body is under cyclic stress (Figure 1.9).

The concept of oxidative stress originated from the human understanding of aging. In the mid-1950s, Professor Harman of the United States put forward the theory of free radical aging for the first time. This theory believes that free radical attacks the living cells and damages them, which is the fundamental cause of aging. It also induces malignant tumors, an important cause of diseases. Professor Sohal, the authoritative professor of aging research in 1990, pointed out the defects of the theory of free radical aging, and was the first to propose the concept of oxidative stress.

Oxidative stress refers to the following circumstance: when the body is in a variety of harmful stimulation, high activity molecules such as ROS and RNS produced too much beyond the removal capability, which might cause the unbalance of oxidative and antioxidant systems, resulting in even further tissue damage.

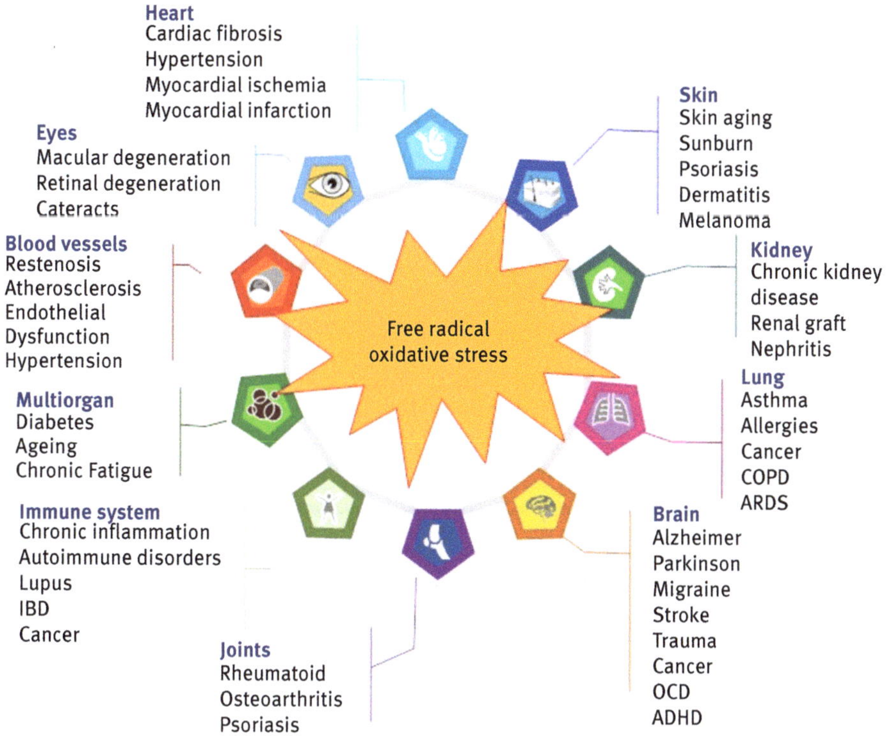

Figure 1.9: Disease caused by free radical imbalance in a human body.

1.3.1 Cardiovascular disease

Cardiovascular disease is a general term for cardiovascular and cerebrovascular diseases. It refers to ischemic or hemorrhagic diseases in the heart, brain, and body tissues caused by hyperlipidemia, blood viscosity, atherosclerosis, and hypertension. Cardiovascular diseases are the major cause of death globally, with 17.5 million people dying of cardiovascular disease in 2012, accounting for 31% of the global total. In the United States, nearly 500,000 people die of myocardial infarction each year [8]. In 1948, in the Framingham Heart Study, 5,209 males and females participated and other important risk factors for cardiovascular disease in addition to dyslipidemia were studied. The results of the study are listed in Table 1.1.

In addition, preliminary results from the Framingham study suggest that high concentrations of total cholesterol and low-density lipoprotein (LDL) cholesterol lead to cardiovascular disease, and the effects of high triglycerides play only a minor role [9]. However, subsequent studies show that elevated triglyceride concentration is an important factor affecting the occurrence of heart diseases. According to the third report of the National Panel of Experts on Cholesterol Education, it is

Table 1.1: Risk factors for cardiovascular disease.

Uncontrollable incentive	Controllable incentive
– Male	– Dyslipidemia (may also be hereditary)
– Senior (male above 45, female above 55 years old)	– hypertension
– Postmenopausal	– Diabetes
– Family history (father or first-degree male relatives died of myocardial infarction or sudden death at age less than 55 years; mother or first-degree female relatives died of myocardial infarction or sudden death at age 65)	– Smoking – Obesity (20% above standard weight), lack of physical activity – Excessive drinking (moderate drinking is beneficial to reduce the incidence of cardiovascular and stroke)
– Genetic factors (Afro-American, Mexican-American, Native American Indian, and South Asian subcontinent have a higher risk of heart disease than Caucasians)	– Bad eating habits (lack of fruit and vegetable intake, high intake of carbohydrates) – Excessive stress response

generally accepted that the ideal total cholesterol content should be less than 200 mg/dL in the body. The results are listed in Table 1.2.

Atherosclerosis is a common cardiovascular disease. Its pathogenesis has been extensively studied by scholars. The mechanism of action by hyperlipidemia and

Table 1.2: Relationship between blood lipid levels and risk of cardiovascular diseases.

Analyte	Cholesterol (mg/dL)	Remarks
Total cholesterol	<200	Ideal state
	200–239	Close to critical value
	>240	High
LDL cholesterol	<100	Best state
	100–129	Good state
	130–159	Close to critical value
High-density lipoprotein	>160	Too high
cholesterol	<40	Low
Triglyceride	≥60	High (appropriate)
	<150	Ideal state
	150–199	Close to critical value
	200–499	High
	>500	Very high

oxidative stress has been established [10]. LDL is the main carrier of cholesterol; LDL is usually present in a nonoxidized state. If endothelial cells dysfunction, xanthine oxidase, lipoxygenase, nicotinamide adenine, and dinucleotide phosphate oxidase [11] will produce excess ROS. ROS further reacts with LDL to form ox-LDL, which phagocytic receptors intake on macrophages without any negative feedback regulation. As a result, a large amount of cholesterol accumulates in phagocytes to form foam cells. Eventually, they will deposit on the coronary artery wall, which not only thicken and harden the arterial wall, but also decline its elasticity. Such a damage will progressively develop into atherosclerotic plaque. On the other hand, ox-LDL not only has cytotoxicity but can also directly damage vascular endothelial cells and smooth muscle cells. This can not only promote monocytes and leukocytes adhesion to the intima, but can also stimulate endothelial cells and can release phagocytic interleukin-1. These can further lead to the occurrence and development of atherosclerosis. However, ox-LDL does not completely explain the complex mechanisms of oxidative stress and atherosclerosis. NO can promote vasodilation, inhibition of platelet aggregation, monocyte adhesion, and LDL oxidation modification, and is also believed to prevent the occurrence of atherosclerosis. Therefore, ox-LDL is involved in the pathogenesis of atherosclerosis in the process of NO inactivation [12]. ROS can directly reduce the NO activity by the reaction of superoxide and NO to produce peroxynitrite salts. At the same time, ROS and peroxynitrite salts oxidize, the cofactor is tetrahydrobiopterin, with NO synthase, as well as block the production of NO [13].

In fact, the underlying cause of atherosclerotic changes in the vascular wall is a chronic inflammatory reaction and immune cytopathic effect.

Ox-LDL is proinflammatory and immunostimulatory, resulting in cells that die at a higher rate. Biologically active phospholipid oxides (ox-PLs) can also be involved in atherosclerosis, which can induce a number of signaling pathways by a different mediating mechanism (other than proinflammatory cytokines), such as tumor necrosis factor or lipopolysaccharide inflammatory gene expression. In addition, oxidized fatty acids on ox-PLs are more hydrophilic and can easily bind to various receptors, which have a significant impact on the inflammatory response and immune response during the host defense response period. Although low levels of oxidized phospholipids act on cells and induce many cytoprotective responses, ox-PLs in focal areas promote atherosclerosis [14]. In addition to the oxidation of low-density lipoprotein and phospholipids, oxidative stress directly activates coronary artery inflammation genes and alters the signal transduction mechanisms of arterial wall cells that regulate the normal function of blood vessels.

Interestingly, however, estrogen in premenopausal women protects them from the detrimental effects of this oxidative stress [15]. The factors that trigger cardiovascular disease in premenopausal women remain to be further studied.

1.3.2 Cancer

Based on the regulation of uncontrolled division and unlimited proliferation of cells, these tumor cells can be divided into benign and malignant. The main difference between benign and malignant tumor cells is that the metastatic tumor is a malignant tumor. In simple terms, malignant cells have reduced intercellular adhesions, thus easily infiltrating surrounding healthy tissues, migrating through the bloodstream or lymphatic pathways, and adhering and proliferating elsewhere.

Cancer refers to malignant tumors that originate in epithelial tissues. Cancer is a major cause of death, and the number of people who die of cancer in the world annually is about 0.01%–0.035% of the total population. These tumors occur as a result of gradual accumulation of genetic mutations. Various environmental agents (such as chemical agents (e.g., chemical mutagenesis), physical factors (e.g., radiation and viral infections), and other biological factors) and intrinsic factors (such as genetic, immune, and endocrine factors) are the carcinogenic factors in a coordinated or sequential manner causing damage of nonlethal cell DNA, thereby activating the protooncogene or tumor suppressor gene mutations. In the induction of cancer initiation, promotion, and development, chronic inflammation is a carcinogenic risk factor, and inflammation is always accompanied by oxidative stress and lipid peroxidation. From this perspective, we can explore the relationship between oxidized active substances and tumors so as to prevent cancer and malignant transformation in the most effective manner.

The types of cancer types caused by oxidative stress are bladder, lung, liver, melanoma, thyroid, colon, oral, ovarian, prostate, breast, kidney, stomach, blood, and pancreatic cancer. For cancer-causing mechanisms, it is generally accepted that excessive oxidative activity in the body attacks biological macromolecules, leading to carcinogenesis [16]. Studies have shown that the hydrogen in methylene between the two double bonds in polyunsaturated fatty acid of biomembrane lipid is active, which is easy to interact with the active oxidation substances to form lipid peroxidation Lipid peroxidation destroys the biomembrane bilayer structure, changing fluidity of the cell membrane. Under aerobic conditions, lipid peroxy radicals will further trigger the chain reaction – the conversion of malondialdehyde, conjugated dienes and hydroperoxides, and other decomposition products. In fact, malondialdehyde, formed by degradation, can also react with free amino groups on molecules such as amino acids, nucleic acids, proteins, and phospholipids to form lipofuscin, which leads to the cross-linking between biological molecules, the replication errors of DNA or inability to split, and the production of amorphous precipitate by proteins. The accumulation of these harmful substances can make the membrane organelles dysfunctional. ROS can also lead to the direct oxidation of proteins, causing protein aggregation, crosslinking, rupturing, and degrading. If a large number of damaged proteins accumulate in the cell, these will induce cell degeneration. Oxidative active substances can also directly attack the DNA strand, resulting in mutations in the protooncogene or tumor suppressor genes in cells, thereby endangering the ability of cells to self-renew and differentiate [17]. In addition, oxidative stress can also destroy intracellular redox balance and can affect some signaling pathways and transcription factors, including Nrf2/Keap1 signaling pathway, nuclear factor kappa B (NF-κB) signaling pathway, activator protein-1, mitogen-activated protein kinases, p53 gene (oncoprotein 53, a tumor suppressor protein), hypoxia inducible factor 1 alpha, and peroxidase proliferator-activated receptor gamma. The activation of these transcription factors affects more than 500 genes involved in the expression of genes such as growth factors, inflammatory cytokines, chemokines, cell cycle control molecules, and anti-inflammatory molecules, suggesting that the oxidative stress is closely related to the induction of cancer contact [18].

1.3.3 Diabetes

Diabetes mellitus is a metabolic disorder characterized by hyperglycemia caused by deficient insulin secretion or impaired insulin function. It is clinically manifest as polyphagia, polyuria, and weight loss. As the criteria for the diagnosis of diabetes, experts from the American Diabetes Association suggest that plasma glucose concentrations should be ≥126 mg/dL after 8 h of fasting or plasma glucose concentrations ≥200 mg/dL at any time of the day under typical hyperglycemic symptoms. With the oral glucose tolerance test (pregnant women, fasting oral glucose 75 g, blood glucose levels after 2 h), the plasma glucose concentration should be ≥200 mg/dL. The aforementioned conditions can be verified on any the other day after diabetes is diagnosed. Continued hyperglycemia and long-term metabolic disorders can lead to chronic complications and can damage the tissues and organs of the body, especially eyes, kidneys, cardiovascular, and nervous system, resulting in several complications. According to Lianhe Zaobao's network report, in 2015, the number of diabetics worldwide is 415 million and will increase by 55% that is to 643 million by 2040, or one out of every 10 adults will have diabetes mellitus Medical expenditures on diabetes will rise from 673 billion to 802 billion dollars, which is an increase of about 20%.

In essence, diabetes can categorized as primary or secondary. At present, most scholars believe that primary diabetes is a genetic disease, and can be divided into insulin-dependent diabetes mellitus (type I diabetes) and noninsulin-dependent diabetes mellitus (type II diabetes). The vast majority of diabetics belongs to this category. Type 1 diabetes often occurs in patients below 30 years, who are prone to ketoacidosis. The etiology is the beta cells present in the pancreas that damage T lymphocyte autoimmune and result in reducing the production of insulin. Therefore, patients need a lifelong external insulin therapy. Type II diabetes accounts for more than 90% of patients with diabetes mellitus; it occurs mainly in people who are more than 40 years. A family history of diabetes or obesity is the major cause of the disease. The pathogenesis is mainly insulin-resistance and insulin-secretion deficiencies. Patients in the early stages of life do not need insulin treatment or can do without ketoacidosis; however, in case of stress, ketoacidosis can be induced. Because of the slow development of hyperglycemia, many patients with no typical symptoms at an early stage are not aware of diabetes for many years. However, they have large blood vessels, microvascular, and neuropathies. Secondary diabetes is a chronic hyperglycemic condition caused by a known primary disease that limits insulin secretion or impairs the physiological role of insulin. Secondary diabetes can be considered as a complication of the primary diabetes; after the primary one is cured, secondary diabetes can also be cured. Some common secondary diabetes are as follows: ① pancreatic diabetes: these are caused due to the pancreas, pancreatic stones, pancreatic cancer, pancreatic excision, and so on, leading to severe pancreatic diseases, which in turn result in insulin secretion and secretion disorder-induced diabetes; ② endocrine diabetes: when insulin antagonist release excessive hormones, such as acromegaly, gigantism is caused by excessive secretion of growth hormones; Cushing's disease is caused by excessive secretion of cortisol; and pheochromocytoma-induced epinephrine can cause diabetes; and ③ iatrogenic diabetes: long-term use of diuretics, glucocorticoids, oral contraceptives, and some antihypertensive drugs can cause impaired glucose tolerance and even diabetes.

A large number of studies have shown that oxidative stress plays an important role in the cause and growth of diabetes. Oxidative stress and diabetes mellitus are mutually induced. These can aggravate each other and can cause chronic complications of diabetes. The increase in metabolic substrate caused by excessive energy intake is a major cause of diabetes. Under the conditions of hyperglycemia and high free fatty acid stimulation, large amounts of ROS and RNS are formed by oxidative phosphorylation of mitochondria, self-oxidation of sugar, NADPH oxidase, lipid peroxidase, and NOS, resulting in an oxidation reaction. Excited oxidative stress plays an important role in increase in diabetes [19]. Oxidative stress can damage the islet β cells, especially the structure of the mitochondria, and can promote the apoptosis of β cells. It can also affect the synthesis and secretion of insulin. A signal transduction pathway indirectly injures islet β cells, such as by activation of NF-κB signaling pathway and triggering β-cell inflammatory response; at the same time, it can affect the pancreatic duodenal homeobox PDX-1 gene expression and activity as well as inhibition of intracellular insulin mRNA production, thereby reducing insulin biosynthesis. In addition, for type 2 diabetes, oxidative stress interferes with the phosphorylation of the insulin receptor and receptor substrate proteins, thus affecting phosphatidylinositol 3-kinase activation, inhibiting adipocyte glucose transporter 4 expression, translocation as well as a series of physiological activities related to insulin signaling leading to the development of insulin resistance [20]. At this point, oxidative stress will in turn further increase the blood sugar, leading to a vicious cycle and aggravate diabetes. In addition, blood glucose fluctuations can increase oxidative stress, resulting in dysfunction of tissues and organs and causing a number of diabetic complications. The specific pathological mechanism [21, 22] is as follows: ① oxidation of active substances can cause oxidative modification of LDL, where ox-LDL cannot be recognized by the LDL receptors by macrophages uptake of foam cells, which promote the formation of atherosclerotic plaque block; ② ROS as an intracellular messenger can activate related redox-sensitive transcription factors such as NF-κB and AP1 as well as can promote adhesion factor, procoagulant tissue factor, endothelin and vascular endothelial growth factor, and monocyte chemotaxis 1 and other protein expressions, resulting in vascular smooth muscle cell migration and proliferation, so that the vascular endothelial function was significantly weaker than the diastolic function, which impaired blood flow regulation and interventional diabetes mellitus disease; ③ oxidation of active substances can promote the formation of advanced glycation end product and can activate the polyol pathway, hexose pathway, and protein kinase C pathway, involved in the process of microvascular and macrovascular complications of diabetes; ④ ROS can inhibit NO synthase activity and can reduce NO synthesis and biological activity, leading to vascular dysfunction, and affecting the endothelial cell antithrombotic properties; ⑤ oxidative stress can increase lipid peroxidation, can decrease Na^+–K^+–ATP enzyme activity, thus slowing the nerve conduction velocity, and can also weaken the generation of nerve growth factor and ciliary neurotrophic factors, affecting the regenerative capacity of damaged nerve fibers; ⑥ peroxynitrite ions can cause protein nitration and can change its molecular biology, resulting in muscle cells, endothelial cells, and other apoptosis; and ⑦ oxidized active substances can break the single-stranded DNA, and it will activate poly-ADP ribose polymerase, further damaging the DNA. Thus, oxidative stress can be considered as an important factor for the prevention and treatment of diabetes and its complications. This is the key to the treatment of diabetes.

1.3.4 Neurodegenerative and other diseases

Neurodegenerative diseases are conditions of neuron loss in the brain and spinal cord. Some of the common examples are Alzheimer's disease, Parkinson's disease, spinal muscular atrophy, and Huntington's disease. Neurons constitute the basic unit of the structure and function of the nervous system. The nervous system can control the sensory function, the function of the effector (such as skeletal muscle, smooth muscle, myocardium, endocrine gland, and exocrine gland), and the function of information integration and storage. Neurons in the brain and spinal cord are generally not regenerated, and so excessive damage can be devastating. The membranes of the central nervous system contain high levels of polyunsaturated fatty acids and have poor antioxidant defenses and are therefore vulnerable to attack by oxidative actives. Thus, with time, neuronal cells gradually accumulate the oxidative damage after mitosis, and so neurodegenerative diseases often occur in the elderly. In addition, the oxidative phosphorylation system in mitochondria will produce free radicals, which may damage mitochondrial DNA, protein and phospholipids. Also the free radicals might react with cytochrome b566, resulting in the decrease of ATP levels. All these circumstances will block the oxidative phosphorylation process. These vicious cycles of energy metabolism and oxidative damage are also one of the important pathways leading to cell death.

In 1956, Professor Denham Harman proposed the theory of free radical aging, believing that organism aging is a result of accumulation of free radicals in the body, and that oxidative active substances, such as free radicals, can induce genetic mutations and the decline of various organs. Aging-related diseases caused by oxidative stress include eye diseases, such as glaucoma, cataracts, and macular degeneration; senile hearing impairment such as senile deafness; musculoskeletal diseases such as osteoarthritis and myotonia; and freckles and senile purpura, nevus, benign tumor, malignant melanoma, squamous cell carcinoma, basal cell carcinoma, athlete's foot, contact dermatitis, ulcer, pruritus, and other skin diseases. On the basis of this understanding, a balanced diet, a reasonable lifestyle, and physical exercise are the fundamental ways to delay aging and also an effective way for reducing diseases.

References

[1] Kunwar A, Priyadarsini K I. Free radicals, oxidative stress and importance of antioxidants in human health. J Med Allied Sci, 2011, 1: 53–60.

[2] Gulcin I. Antioxidant activity of food constituents: An overview. Arch Toxicol, 2012, 86: 345–91.

[3] Powere S K, Jackson M J. Exercise induced oxidative stress: Cellular mechanism and impact on muscle force production. Physiol Rev, 2008, 88: 1243–76.

[4] Choudhari K S, Chaudhary M, Bagde S, et al. Nitric oxide and cancer: A review. World J Surg Oncol, 2013, 11: 118.

[5] Al-Ghamdi S S, Rafter M J, Yaqoob M M. Acute solvent exposure induced activation of cytochrome P4502E1 causes proximal tubular cell necrosis by oxidative stress. Toxicol In Vitro, 2003, 17: 335–41.

[6] Valko M, Leibfritz D, Moncol J, et al. Free radicals and antioxidants in normal physiological functions and human diseases. Int J Biochem Cell Biol, 2007, 39: 44–84.

[7] Sena L A, Chandel N S. Physiological roles of mitochondrial reactive oxygen species. Mol Cell, 2012, 48: 158–67.

[8] Movahed M R, John J, Hashemzadeh M. Mortality trends in non-ST-segment elevation myocardial infarction (NSTEMI) in the United States from 1998 to 2004. Clin Cardiol, 2011, 34: 689–92.

[9] Gordon T, Kannel W B, Castelli W P. Lipoproteins, cardiovascular disease and death: The Framingham study. Arch Intern Med, 1981, 141: 1128–31.

[10] Peluso I, Morabito G, Urban L, et al. Oxidative stress in atherosclerosis development: The central role of LDL and oxidative burst. Endocr Metab Immune Disord Drug Targets, 2012, 12: 351–60.

[11] Victor V M, Apostolova N, Herance R, et al. Oxidative stress and mitochondrial dysfunction in atherosclerosis: Mitochondria-targeted antioxidants as potential therapy. Curr Med Chem, 2009, 16: 4654–67.
[12] Victor V M, Rocha M, Sola E, et al. Oxidative stress endothelial dysfunction and atherosclerosis. Curr Pharm Des, 2009, 15: 2988–3002.
[13] Landmesser U, Harrison D G. Oxidant stress as a marker for cardiovascular events: Ox marks the spot. Circulation, 2001, 104: 2638–40.
[14] Lee S, Birukov K G, Romanoski C E, et al. Role of phospholipid oxidation products in atherosclerosis. Circ Res, 2012, 111: 778–99.
[15] Kondo T, Hirose M, Kageyama K. Roles of oxidative stress and redox regulation in atherosclerosis. J Atheroscler Thromb, 2009, 16: 523–38.
[16] Perse M. Oxidative stress in the pathogenesis of colorectal cancer: Cause or consequence? Biomed Res Int, 2013, 2013: 725710.
[17] Cejas P, Casado E, Belda-Iniesta C, et al. Implications of oxidative stress and cell membrane lipid peroxidation in human cancer. Cancer Causes Control, 2004, 15: 707–19.
[18] Reuter S, Gupta S C, Chaturvedi M M, et al. Oxidative stress, inflammation and cancer: how are they are linked? Free Rad Biol Med, 2010, 49: 1603–16.
[19] Yang H, Jin X, Kei Lam C W, et al. Oxidative stress and diabetes mellitus. Clin Chem Lab Med, 2011, 49: 1773–82.
[20] Marshak S, Leibowitz G, Bertuzzi F, et al. Impaired beta-cell functions induced by chronic exposure of cultured human pancreatic islets to high glucose. Diabetes, 1999, 48: 1230–6.
[21] Johansen J S, Harris A K, Rychly D J, et al. Oxidative stress and the use of antioxidants in diabetes: linking basic science to clinical practice. Cardiovasc Diabetol, 2005, 4: 5–15.
[22] Maritim A C, Sanden R A, Watkins J B 3rd. Diabetes, oxidative stress, and antioxidants: a review. J Biochem Mol Toxicol, 2003, 17: 4–38.

2 Endogenous and exogenous antioxidants

During physiological metabolism, there are some highly active oxidative substances in an organism, including ROS and RNS, but none of these cause any serious damage to the organism. Antioxidants are compound that help protect organisms from oxidation of active substances (Figure 2.1). An ideal antioxidant must have the following properties: ① can act on the lesion site in time with appropriate concentration; ② can react with active oxidation substances promptly; ③ the reaction product has decreased toxicity than the original active oxidation species; ④ under certain conditions, antioxidants can restore the initial state for continuous scavenging free radicals. Based on the sources of antioxidants, they can be divided into endogenous [1, 2] and exogenous antioxidants [3, 4].

2.1 Endogenous antioxidants

Endogenous antioxidants are the ones that can produce their own antioxidant substances. Based on their structural features, they can be further divided into enzyme and nonenzyme antioxidants (Figure 2.2).

2.1.1 Enzymes antioxidants

In the process of evolution, human beings evolved with an instinct to eliminate the free radical damage and create antioxidant enzymes to reduce the destruction of free radicals in the cells of the body. Antioxidant enzymes are the first line of defense against the oxidative stress in the body (Figure 2.3).

Enzyme antioxidants play an important part in the body's defense against oxidative stress, including superoxide dismutase (SOD), catalase (CAT), glutathione peroxidase (GPx), glutathione reductase (GR), and so on.

SOD is widely found in animals, plants, and microorganisms and can specifically eliminate superoxide anion and can convert it to H_2O_2 and O_2. Depending on the type of SOD present in a metal, it can be divided into Cu Zn-SOD, Mn-SOD, and Fe-SOD. Cu Zn-SOD is the most common one, which is green and is mainly found in the cytoplasm of eukaryotes; Mn-SOD is purple and is mainly found in eukaryotes mitochondria and prokaryotic cells; and Fe-SOD is brown and is mainly present in prokaryotic cells. The concentration of different kinds of SOD should be controlled by the corresponding gene [5]. If there is SOD gene disorder, the body cannot be cleared of free radicals and will suffer from many diseases. The performance of SOD varies from person to person and also declines with age. In addition, the living conditions change the performance of the gene.

https://doi.org/10.1515/9783110573763-002

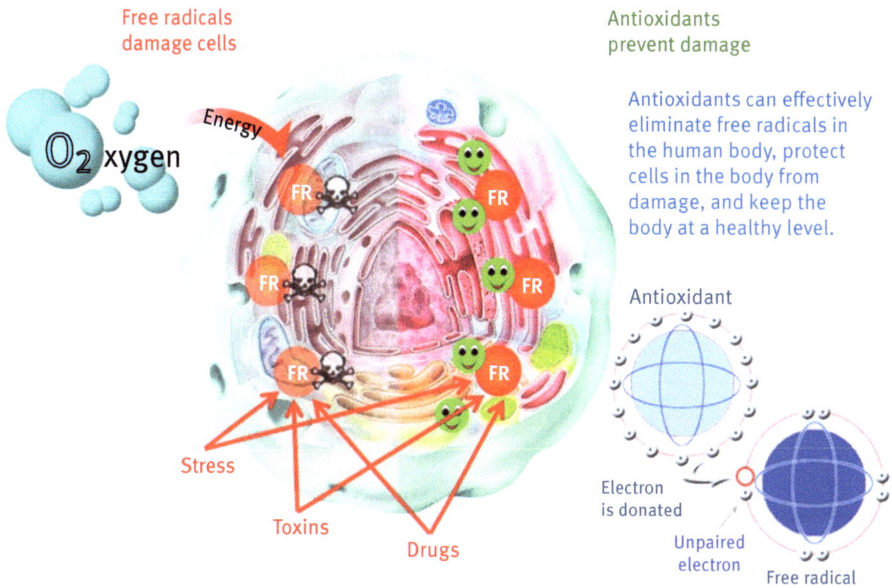

Figure 2.1: Oxidation and antioxidation in a human body.

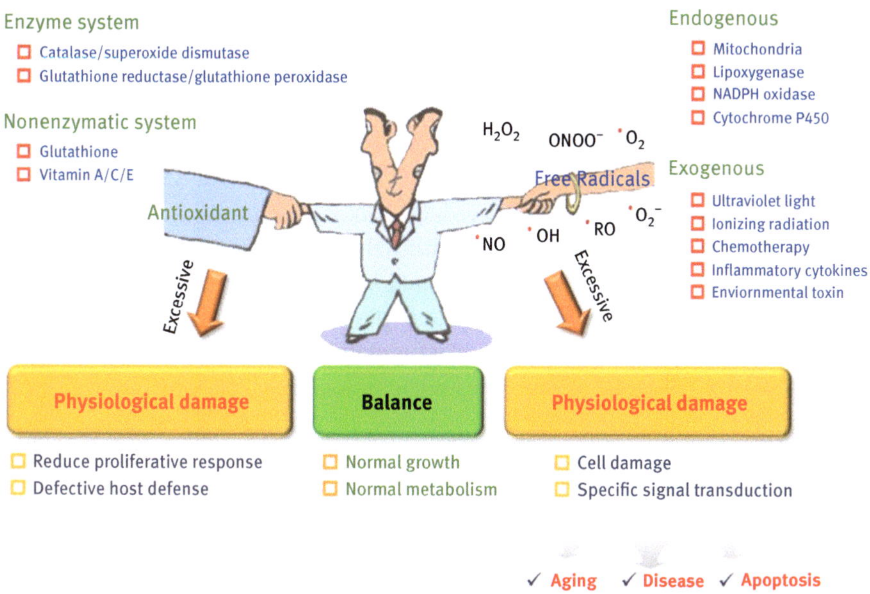

Figure 2.2: Endogenous antioxidant maintains free radical balance in a human body.

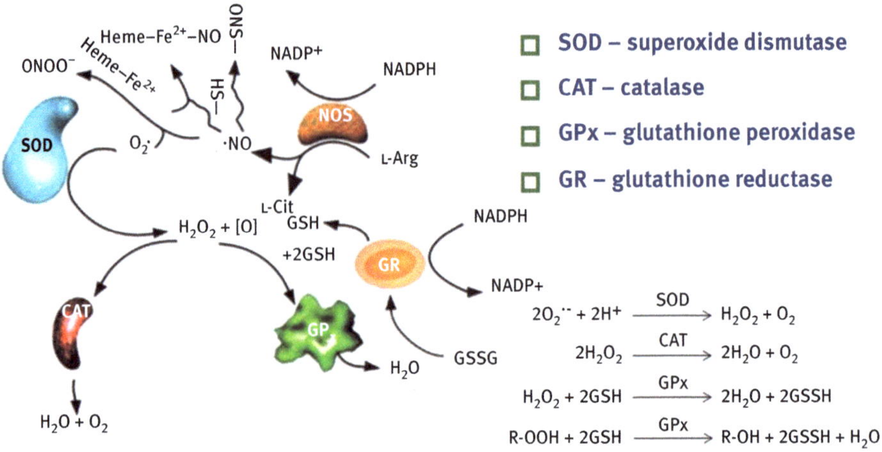

Figure 2.3: Antioxidant action in the body – antioxidant enzyme protection system.

CAT is one of the key enzymes in a biological defense system, where iron porphyrin acts as a cofactor [6], which is distributed in the body's peroxisomes, especially in the liver. When two molecules of H_2O_2 successively collide with the active center of CAT, it can promote the disproportionation reaction of H_2O_2 in the body and transform it into O_2 and H_2O. The higher is the concentration of H_2O_2, the faster is the rate of decomposition.

Based on the fact whether the structure has selenium atoms, GPx can be divided into selenium-containing GPx and nonselenium-containing GPx. The former is a catalytic reduction of H_2O_2 and organic peroxides, which decompose only organic peroxides. GPx, which is one of the main components of the enzymatic protection system that inhibits membrane phospholipid peroxidation, can eliminate superoxide anion and H_2O_2 along with SOD and CAT to reduce or block the primary reaction of lipid peroxidation. At the same time, GPx can also reduce or prevent the secondary reaction by reducing organic peroxide, which would in turn protect the biofilm.

GR is extensively found in animal and plant tissues, microorganisms, and yeasts. It is capable of converting oxidized glutathione to reduced glutathione in the presence of NADPH, while the reduced glutathione may continue elimination of free radicals.

2.1.2 Nonenzymatic antioxidants

Nonenzymatic antioxidants are biologically small molecules that have antioxidant effects. These include vitamin E, taurine, epinephrine, ascorbate, glutathione, β-carotene, melatonin, uric acid, lipoic acid, ubiquinone/Co Q10, metal selenium, and zinc.

Vitamin E is a fat-soluble vitamin. Its hydrolyzate is alpha-tocopherol, which is the main antioxidant on the cell membrane and can inhibit the unsaturated lipid peroxidation chain reaction. The oxidation of the product can be reduction of vitamin C.

Ascorbic acid, also known as vitamin C, can effectively remove superoxide radicals, hydroxyl radicals, and hypohalous acids. It can end the free radical chain reaction and inhibit the generation of free radicals. In the human body, hydrogen atom from ascorbic acid molecule should be passed to vitamin E radical, which will further lead to the regeneration of vitamin E and thus realize its indirect function as active antioxidant substance. Also, vitamin C is likely to perform the Synergy together with α-tocopherol or reduced coenzyme II to eliminate free radicals in human body.

Glutathione is a thiol-containing tripeptide composed of glutamic acid, cysteine, and glycine. Under physiological conditions, it has oxidation and reduction of two forms. As an important antioxidant in the body, it can protect many sulfhydryl groups in proteins and enzymes [7]. It can also protect hemoglobin from transporting oxygen to be oxidized by H_2O_2, free radicals and the like.

As shown in Figure 2.4, vitamins C and E, glutathione, and other hydroxyl radicals can be directly produced by the generation of inactive radical products. In addition, vitamin C can restore the roles of free radicals that were produced by vitamin E and can also restore original antioxidant functions; finally, vitamin C is converted to the excreted oxalic acid .

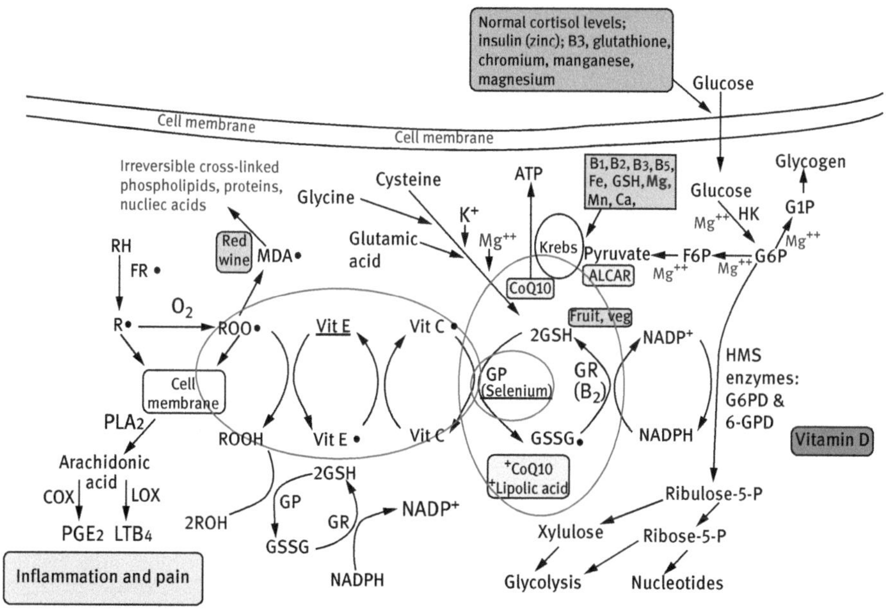

Figure 2.4: Examples of cellular antioxidant processes involved in vitamin C/E.

The chemical name of taurine is 2-aminoethanesulfonic acid, which is also known as bile acid and bovine choline. This is a sulfur-containing nonprotein amino acid, which exists in the free state and is not involved in protein biosynthesis. This can increase the scavenging effect on the oxidized active substances, can stabilize the cell membrane, can reduce the formation of lipid peroxidase and the leakage of myocardial enzymes, and can decrease the apoptosis rate of cardiomyocytes by resisting the damage of the exogenous hydroxyl radicals to the cardiomyocytes. It can also improve the immunity of skin by weakening the effects of external environment on the skin.

Adrenaline is a catecholamine hormone and neurotransmitter secreted by the adrenal gland that promotes the breakdown of glycogen.

β-Carotene is a precursor of vitamin A, which can effectively capture the body of reactive oxygen species, though when combined with lipid peroxidation free radicals its peroxide chain reaction is interrupted. However, vitamin A is not as good as β-carotene antioxidant; overdose of vitamin A can be poisonous, causing nausea, hair loss, bones, soreness, sleepiness, and so on. In case of pregnant women, excessive use of this may also lead to fetal malformations. β-Carotene in the human body has two functions as antioxidants: ① this can be combined with lipid peroxidation free radicals and can lead to lipid peroxidation chain reaction and ② can absorb oxygen when exposed to light (as in the eyes) and can result in enough energy to stop the oxidation reaction. However, excessive intake of β-carotene is not found to cause danger of poisoning.

Melatonin is an amine hormone produced by mammals or pineal glands in the body, resulting in whitening of the skin. It can easily penetrate the cell membrane and blood–brain barrier, but does not participate in the reduction cycle. This means that after oxidation it is not renewable; its oxidized melatonin will form a stable and free radical end product and will have a protective cell structure. To prevent DNA damage, the role of peroxide content, the so-called terminal antioxidants, in the body should be regulated [8].

Uric acid, a byproduct of purine metabolism, is a major metabolite in birds and reptiles, and is present in small amounts in normal human urine. It has the highest concentration of biological antioxidants in the blood. Although it cannot react with superoxide, it can scavenge peroxynitrite, peroxide, and hypochlorous acid.

Lipoic acid, 6,8-dithioic acid, is a coenzyme present in mitochondria that functions like a vitamin and is also involved in acyl transfer in metabolism, acting as a hydrogen and acyl transmitting agent. Because it is both fat- and water-soluble, is significantly electrophilic, and can react with free radicals, it is known as the "universal antioxidant."

Ubiquinone, also known as coenzyme Q (Co Q), is a fat-soluble antioxidant. The structure is similar to that of vitamins K and E and plastoquinone with side chains consisting of 6–10 isoprene units, whereas mammalian cells contain 10 isoprene units and are therefore called Co Q10 [9]. Ubiquinone, which is mostly present in the mitochondria of eukaryotic cells, is a hydrogen transporter in the respiratory chain, that is, hydrogen is obtained by the redox reaction and is transferred to cytochromes to obtain the electron transfer function.

Metallic selenium [10] and zinc, although do have no antioxidant effect, play a crucial role in the activity of antioxidant enzymes and are considered as antioxidant nutrients.

Some transferrin, iron-binding hemoglobin, copper-binding serum albumin, and so on can chelate the transition metal and can also chelate the tissue metal to catalyze the generation of free radicals in the cell and can help in antioxidant defense.

In vivo antioxidants are shown in Figure 2.5.

2.2 Exogenous antioxidants

Exogenous antioxidants are those that are taken in vitro. Owing to aging, diseases, and other external environmental factors, there are genetic changes, functioning of a body declines, and immune functions significantly decrease, thereby reducing the ability to resist oxidative stress and body antioxidant deficiencies. On such cases, the external supplement intake of antioxidants can provide additional protection.

Commonly found antioxidants in foods are vitamins, terpene-carotenoids, hormones, flavonoids, phenolic acids and their esters, non-flavonoid phenols and other organic classes; these are listed in Table 2.1.

Vitamin is a kind of trace organic substance, which is a must for the body to maintain normal physiological functions, and it plays an important role in growth, metabolism, and overall development. We normally intake vitamins from food. Vitamins neither contribute to the formation of cells nor provide energy to the body like sugars, proteins, and fat, but they play a significant role in regulating the metabolism of the organism. Based on solubility classification, vitamins can be grouped into two types: ① fat-soluble vitamins, including vitamins A, D, E, and K and ② water-soluble vitamins, including vitamins B1, B2, B3, B5, B6, B7, B9, B12, and vitamin C. Among these, only vitamins A, C, and E act as an antioxidants. We consume vitamin C mainly from oranges, grapefruit, kiwi, strawberry, cantaloupe, red pepper, green pepper, Brussels sprouts, broccoli, tomatoes, cabbage, spinach, and peas. High contents of vitamin E are obtained from oatmeal, spinach, asparagus, broccoli, cucumber, avocado, hazelnut, peanut butter, almonds, vegetable oil, eggs, meat, and poultry, while high levels of vitamin A are obtained from liver, fish, eggs, milk, ice cream, cheese, and cereals. We get β-carotene from spinach, carrots, cauliflower, cantaloupe, sweet potato, pumpkin, mango, apricot, nuts, and cowpeas, which gets converted to vitamin A after metabolizing and is thus indirectly related to the production of vitamin A .

Carotenoids are terpenoids in four terpene compounds: ubiquitous in animals, advanced plants, fungi, and algae in the yellow, orange, or red pigment. They have a

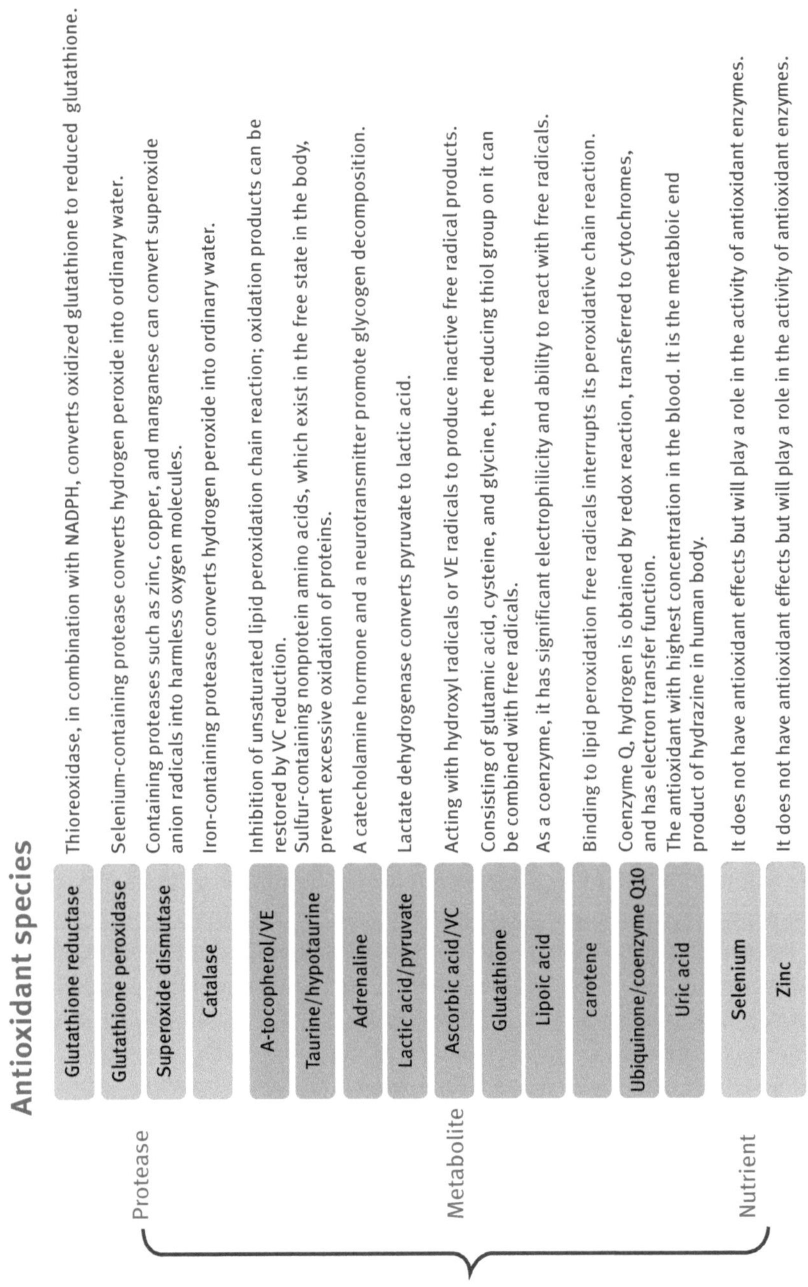

Figure 2.5: Antioxidant species in a human body.

Table 2.1: Common antioxidants in food.

Classification	Species
Vitamins	Vitamin A (retinol/β-carotene), vitamin C, vitamin E (tocotrienol/tocopherol)
Terpenoids-carotenoids	α-Carotene, β-carotene, astaxanthin, canthaxanthin, lutein, lycopene, zeaxanthin
Hormone	Melatonin
Flavonoids	Flavonoids (apigenin/luteolin/red orange), flavonol (isoordin/kaempferol/myricetin/proanthocyanidin/quercetin), flavanone (spirulina/hesperet/pomelo), flavonoids and their polymers (catechin/gallocatechin/epicatechin/epigallocatechin-lipid/aflavin-lipid/chelucin); isoflavone phytoestrogens (daidzein/genistein/daidzein), stilbene (resveratrol/rosewood), anthocyanin (anthocyanin/delphinidin/dimethylformin/geranium pigment/methyl anthocyanin/petunidin)
Phenolic acid and its esters	Cichoric acid, chlorogenic acid, caffeic acid, ferulic acid, cinnamic acid, ellagic acid, ellagitannin; gallic acid, gallic acid tannin, rosmarinic acid, salicylic acid
Nonflavonoid phenols	Curcumin, flavonoid lignin, anthrone, eugenol
Other organics	Capsaicin, bilirubin, citric acid/oxalic acid/phytic acid, acetylcysteine, uric acid

series of conjugated double bonds and methyl branches. The more the number of conjugated double bonds, the finer is the color of the pigment (from red). Because of the presence of double bonds, carotenoids act as singlet oxygen quenchers, which protect the body from damage.

Hormones are drugs that are active ingredients of human and animal hormones (including organic compounds that work in the same way as hormonal structures and functions). Melatonin is a hormone secreted by the pineal gland, which promotes good sleep, adjusts the time difference, acts as antiaging agent, and regulates immune system, antitumor, and other physiological effects. In general, with aging there is not enough melatonin secretion, thus affecting sleep; hence, the intake of melatonin tablets for proper supplementation becomes important.

Flavonoids are secondary metabolites of plants and exist in the form of flavonoid glycosides or flavonoid glycosides in food-borne plants such as fruits, vegetables, beans, and tea. Flavonoids have strong antioxidant properties, can be used as metal chelators and reducing agents as well as reactive oxygen species elimination agents, and help terminating the free radical chain reactions and singlet oxygen formation. Different types of flavonoids have different degrees of antioxidants. The number and location of hydroxyl substituents have a considerable impact on antioxidant effects. The more the number of hydroxyl groups, the greater is the ability to scavenge free radicals.

Phenolic acids refer to a class of organic acids containing several phenolic hydroxyl groups on the same benzene ring. The phenolic hydroxyl groups on the benzene rings of these materials are groups with antioxidant activities and mainly rely on the dehydrogenation reaction to obtain stable phenoxy free radicals so as to stop the chain reactions. Foods obtained from plants are rich in phenolic acids and esters, for example, caffeic acid, ferulic acid, and chlorogenic acid.

Curcumin is a yellow pigment extracted from rhizomes of Zingiberaceae, Araceae, and so on. Its main chain is an unsaturated aliphatic and aromatic group and is a dione compound. Owing to the presence of o-methylated phenols, which belong to polyphenols, with properties like lipid-lowering, antioxidation, anti-inflammatory, antiatherosclerosis, and other medicinal values (though not in the structure of flavonoids), it is classified as nonflavonoids phenols.

Phytic acid, also known as phytic acid, is a class of organophosphates that contain the highest levels of phosphorus in legumes, cereals, and bran, and is the major form of phosphorus stored in plants. Phytic acid can provide six pairs of hydrogen atoms, forming a stable structure with free radicals, with food preservation effects. The hydrolyzates of phytic acid are inositol and phospholipids. The former has antiaging effect to some extent and the latter is an important component of human cells. However according to the molecular structure of phytic acid, it cannot be classified to any of the above-mentioned categories. Hence, it is considered as other organic antioxidant, which will realize its effect.

2.3 Antioxidant intake

Owing to aging, diseases, and external environmental factors, excess free radicals lead to changes in genes, body functions decline, and immune functions weaken, resulting in reducing the ability to resist oxidative stress and decreasing antioxidants in the body. Although there is a mechanism of antioxidant protection in the body, some free radicals still escape the protection system and cause oxidative damage. In such cases, dietary intake of AO can play an additional protective role.

Although antioxidant enzymes are the first line of defense against the effects of antioxidants in the body, it is difficult to orally supplement antioxidases and these proteins break down into small molecules, such as amino acids via gastric acid.

In the usual natural diets, the three major antioxidants are vitamin E, vitamin C, and β-carotene. In addition, some of the phytochemicals, such as flavonoids, indoles, Genistein, lycopene, and so on, also have antioxidant effects.

Some water-soluble antioxidants, such as vitamin C, can easily be excreted and cannot be stored, and so there is a daily need of external supplements.

Iron, copper, zinc, and selenium are all essential ingredients in the production of antioxidant enzymes. Although not essential, they must be present in the body to replenish antioxidant enzymes, which we intake usually from the diet.

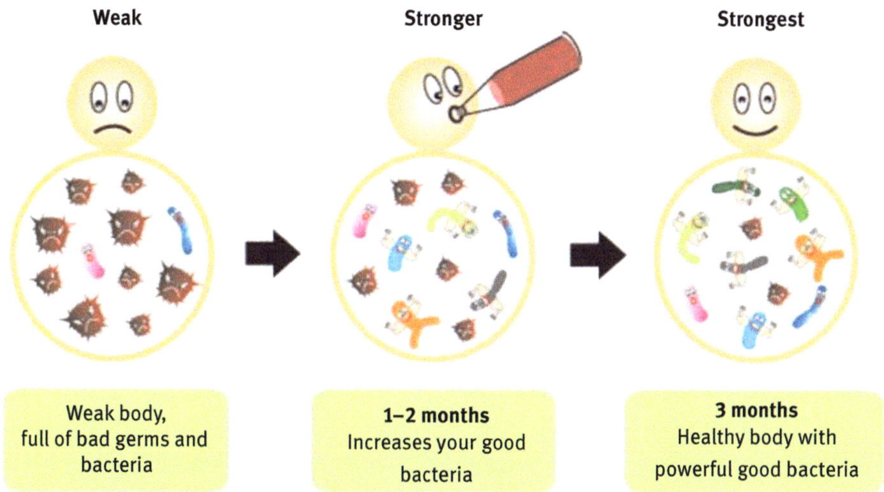

Figure 2.6: Intake of Antioxidants and health.

The health of the body depends to a good extent on the intake of antioxidants. In recent years, there are about 20% of papers published on this topic in *Journal of Agriculture and Food. Chemistry*. Each issue is someway related to the analysis of food antioxidants and free radical damage (Figure 2.6).

References

[1] Levine R L, Mosoni L, Berlett B S, et al. Methionine residues as endogenous antioxidants in proteins. Proc Natl Acad Sci USA, 1996, 93: 15036–15040.

[2] Schreibelt G, Horssen J V. Therapeutic potential and biological role of endogenous antioxidant enzymes in multiple sclerosis pathology. Brain Res Rev, 2007, 56: 322–330.

[3] Fernandez-Panchon M S, Villano D, Troncoso A M, et al. Antioxidant activity of phenolic compounds: From in vitro results to in vivo evidence. Critical Reviews in Food Science and Nutrition, 2008, 48: 649–671.

[4] Grimble R F. Nutritional antioxidants and the modulation of inflammation: Theory and practice. New Horizons, 1994, 2: 175–85.

[5] Miao L, Clair D K S. Regulation of superoxide dismutase genes: Implications in disease. Free Rad Biol Med, 2009, 47: 344–56.

[6] Kirkman H N, Gaetani G F. Mammalian catalase: A venerable enzyme with new mysteries. Trends Biochem Sci, 2007, 32: 44–50.

[7] Droge W. Aging-related changes in the thiol/disulfide redox state: Implications for the use of thiol antioxidants. Experimental Gerontology, 2002, 37: 1333–1345.

[8] Rodriguez C, Mayo J C, Sainz R M, et al. Regulation of antioxidant enzymes: A significant role for melatonin. Journal of Pineal Research, 2004, 36: 1–9.

[9] Eroglu M, Sahin S, Durukan B, et al. Blood serum and seminal plasma selenium, total antioxidant capacity and coenzyme q10 levels in relation to semen parameters in men with idiopathic infertility. Biological Trace Element Research, 2014, 159: 46–51.

[10] Refat M S, Kobeasy M I, El-Shazly M A. Synthesis, spectroscopic, and molar conductance characterization of selenium(IV)vitamin B6 complex as prospective antioxidant agent. Rus J General Chem, 2014, 84: 1411–1416.

3 Determination of antioxidant capacity by optical methods

On the basis of Lambert–Beer's law, the optical colorimetric method determines contents of a component under study by comparing or measuring the depth of the color of the colored substance solution. This method is simple, quick, and quantitatively measurable. Visual colorimetry or photoelectric colorimetry is widely used in the measurement of antioxidant capacity. In addition, in the recent years, some fluorescence analysis and spectroscopic methods have also been applied in the field of antioxidant measurement.

Because of the complexity of antioxidants, in foods it is difficult to separate and analyze each component one by one. In addition, there are still some synergistic effects among the antioxidant components. Therefore, a benchmark needs to be set to measure the relative datum – the amount of the sample. In general, a water-soluble vitamin E mimics Trolox (6-hydroxy-2,5,7,8-tetramethylchroman-2-carboxylic acid), which acts as the reference. Trolox equivalents (TE) are commonly used as units of measurement, expressed as μmol TE/100 g. The molecular structure of Trolox is shown in Figure 3.1.

The optical method is mainly based on the concept of the change in color or luminescence before and after the reaction of the colored reference material and the oxidized active material as well as the influence of the addition of the antioxidant on it. Based on different measurement principles, the optical method can be divided into hydrogen atom transfer (HAT) method and single-electron transfer (SET) method [1].

3.1 HAT method

3.1.1 Oxygen-free radical absorption capacity of the method

The oxygen radical absorbance capacity (ORAC) method is based on the antioxidant oxidative damage caused by the inhibition of peroxy radicals, which shows that when peroxyl radicals react with the fluorescent probe they generate a nonfluorescent composite. On the basis of the fluorescent probe used for the measurement [2–5], the ORAC method for the determination of the process is shown in Figure 3.2.

The ORAC assay can be used for simultaneously detecting the antioxidant capacity of water- and fat-soluble antioxidants; however, the major drawback is the longer analysis time (about 1 h). Following are the equations for the reaction process and testing methods:

https://doi.org/10.1515/9783110573763-003

$$R-N=N-R \xrightarrow{O_2} N_2 + 2ROO\cdot$$

$$ROO\cdot + AH \longrightarrow ROOH + A\cdot$$

$$ROO\cdot + A\cdot \xrightarrow{\text{fast}} ROOH$$

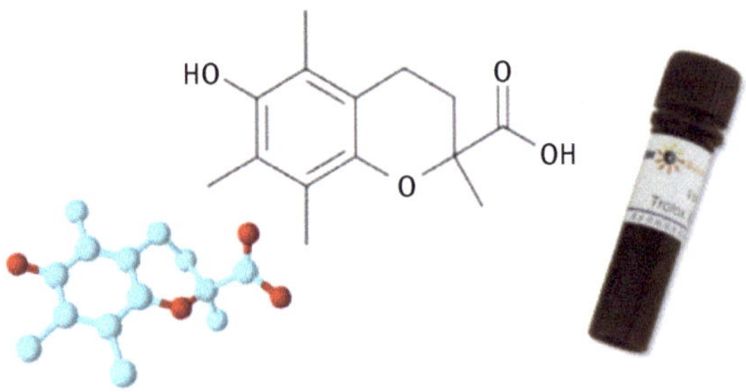

Figure 3.1: The molecular structure of Trolox.

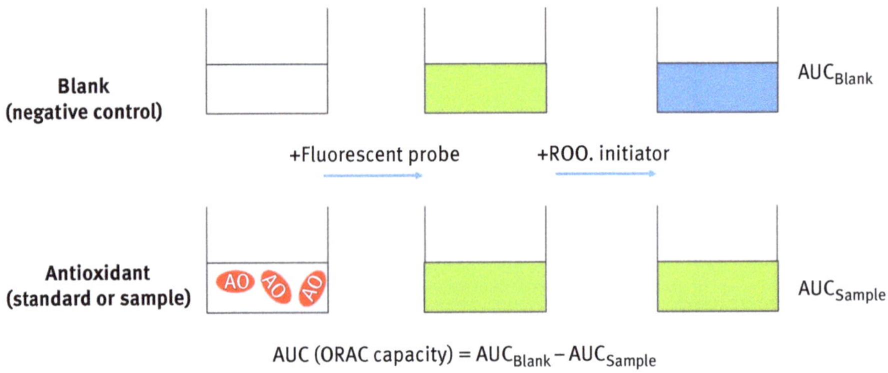

Figure 3.2: Detection of ORAC.

3.1.2 Total free radical trapping antioxidant parameter method

The determination principle of the total free radical trapping antioxidant parameter method [6, 7] is based on the fact that 2,2′-azo (2-methylpropionamidine) pyrolysis consumes oxygen to produce peroxy radicals [8]. The antioxidant capacity was tested by measuring the oxygen consumed. This method is similar to the ORAC method, but the stability of the oxygen electrode limits its use. Researchers have made some modifications to this method. Using fluorescent probes (such as R-phycoerythrin) [8, 9] instead of the oxygen ones has helped to overcome the shortcomings to a certain extent.

3.1.3 Photochemiluminescence method

Superoxide radicals are commonly used in photochemiluminescence (PCL) method, where the PCL behavior is finally studied for antioxidant capacity evaluation. The specific process is shown in Figure 3.3, where S is a photosensor.

$$S + h\nu + O_2 \longrightarrow [S^{*}O_2] \longrightarrow S^{*} + O_2^{-}$$

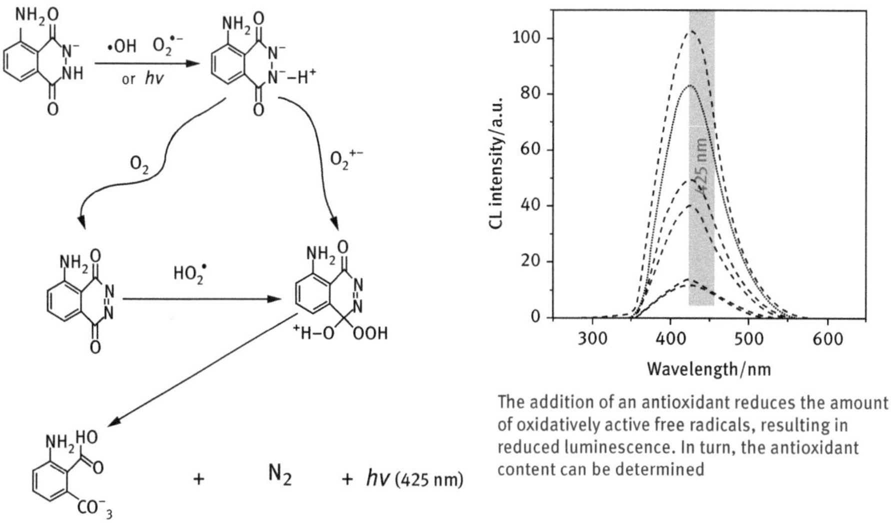

Figure 3.3: Principle of photochemical luminescence detection.

In the absence of an antioxidant, a higher concentration of the oxidatively active material oxidizes the luminescent material to render it a stronger luminescent property. The addition of the antioxidant also lowers the contents of oxidatively active free radicals, thereby decreasing luminescence. The intensity of light can be determined based on the antioxidant capacity of antioxidants [10–12]. According to Popov and Lewin, Jena Germany designed a photochemical luminescence antioxidant capacity detector called Analytik Jena AG; the instrument has been widely used for the determination of antioxidant capacity. PCL strategy is available for the assay of both the fat- and water-soluble antioxidants. The major disadvantage of this method is that only one sample can be measured at a time. Thus, this method cannot achieve high-throughput determination.

3.1.4 LOO· bleaching β-carotene reaction assay

β-carotene is easily oxidized by LOO· and causes the color fading. The introduction of antioxidants can protect β-carotene from being oxidized. Based on this principle, the antioxidant capacity of antioxidants can be measured [13–15]. The method is performed at 443 nm [16], and so it does not require any special or complex instruments. This method can also measure the antioxidant capacity of high-throughput antioxidants.

Other methods for determining hydrogen transfer include total oxidant abatement [17–20], chemiluminescence [21–25], and low-density lipoprotein oxidation [26–29].

3.2 SET reaction

3.2.1 Determination of capacity of AO-reduced iron (Fe (III))

The reduced state of Fe (III) $(TPTZ)_2Cl_3$ (2,4,6-tripyridyl-triazine, tripyridyltriazine, TPTZ) has a strong optical absorption at 593 nm. When an antioxidant is added, the reduced state of $[Fe (III) (TPTZ)_2]^{2+}$ increases (Figure 3.4). Therefore, based on the optical absorption value and kinetics, the antioxidant activity can be determined [30–34]. Advantages of AO-reduced iron (Fe (III)) capacity (FRAP) method are as follows: it is simple and fast, cheap, extensive, automatic and semiautomatic measurement without any special equipment, and can be used for polyphenols containing foods, beverages,

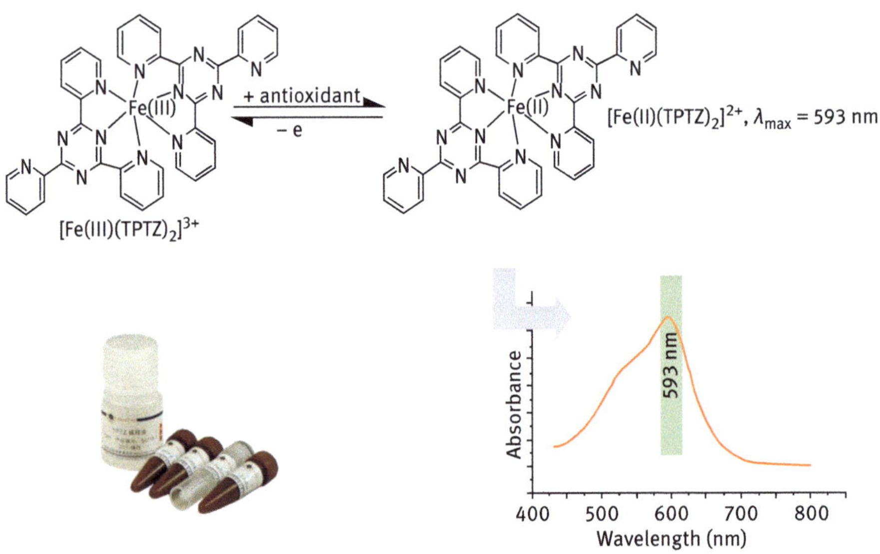

Figure 3.4: The principle for the determination of antioxidant capacity by FRAP spectroscopy.

and nutritional products for the analysis of antioxidant capacity. The disadvantages are the determination takes a longer time, thiols antioxidants cannot be measured, and some of the carbohydrates present can affect the measurement results.

3.2.2 Copper (Cu (II)) reduction colorimetric method

The bathocuproine chromatoxyl colorimetric assay is a new method developed by Apak et al. [35–39]. According to the system color change caused by different valence states of Cu (II, I) complex, antioxidant capacity is determined (Figure 3.5). When an antioxidant is added, the concentration of Cu (I) complex increases. Then, the absorbance of Cu (I) complex at 490 nm can be measured and the antioxidant capacity of antioxidant can be determined. The principles of the CUPRAC and FRAP methods are basically similar. Cu (II) when compared with Fe (III) has a low redox potential. Using CUPRAC method for determining the antioxidant capacity can effectively reduce the interference of carbohydrates, and the measurement time is also less. However, Cu (II) can also measure thiol antioxidants [1]. The CUPRAC method has its own drawbacks: it takes a longer time to measure mixed antioxidants, and the resulting products are less

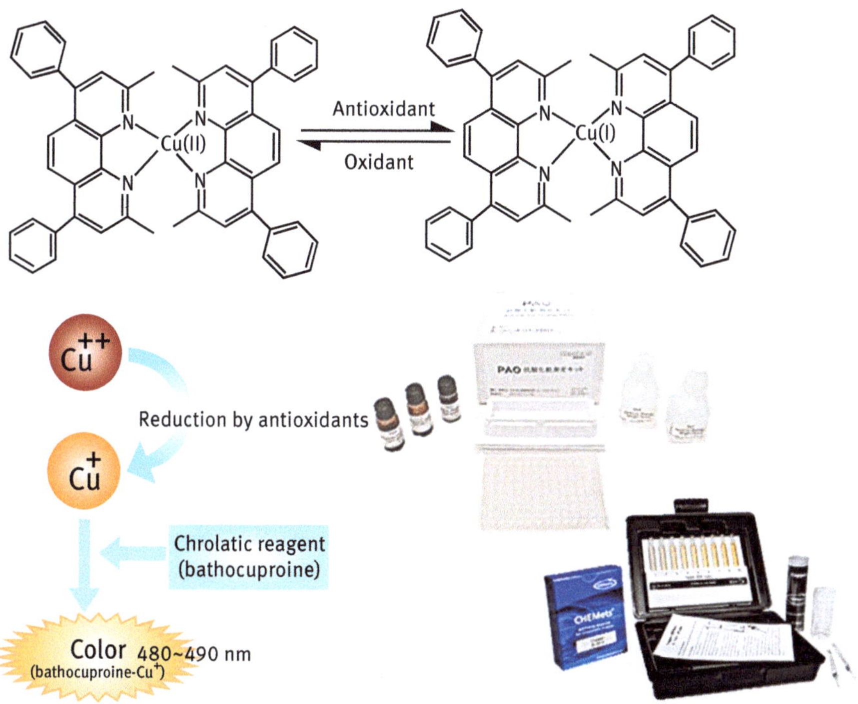

Figure 3.5: Determination of antioxidant capacity by copper ion reduction method.

stable due to the spatial conformation effect. Nowadays, kit products, based on the CUPRAC method for determining the antioxidant capacity, such as bathocuproine (2,9-dimethyl-4,7-diphenyl-1,10-phenanthroline), are commercially available.

3.2.3 Dibenzyl hydrazide free radical quenching method

DPPH (516 nm) has optical absorption properties and can be used both as a free radical and as a reaction indicator in free radical quenching. The decrease in DPPH absorption is proportional to the antioxidant activity [40–45] (Figure 3.6). Recent studies show that DPPH has both SET and HAT mixing mechanisms. The DPPH spectroscopic method is suitable for the analysis of antioxidant activity of medicinal plants. DPPH method helps determine the wavelength even at 516 nm; however, the overlap of wavelengths and colored substances present in foodstuffs pose challenges for the DPPH method. DPPH can only be dissolved in organic solvents, such as ethanol, which is not suitable for the detection and analysis of water-soluble samples.

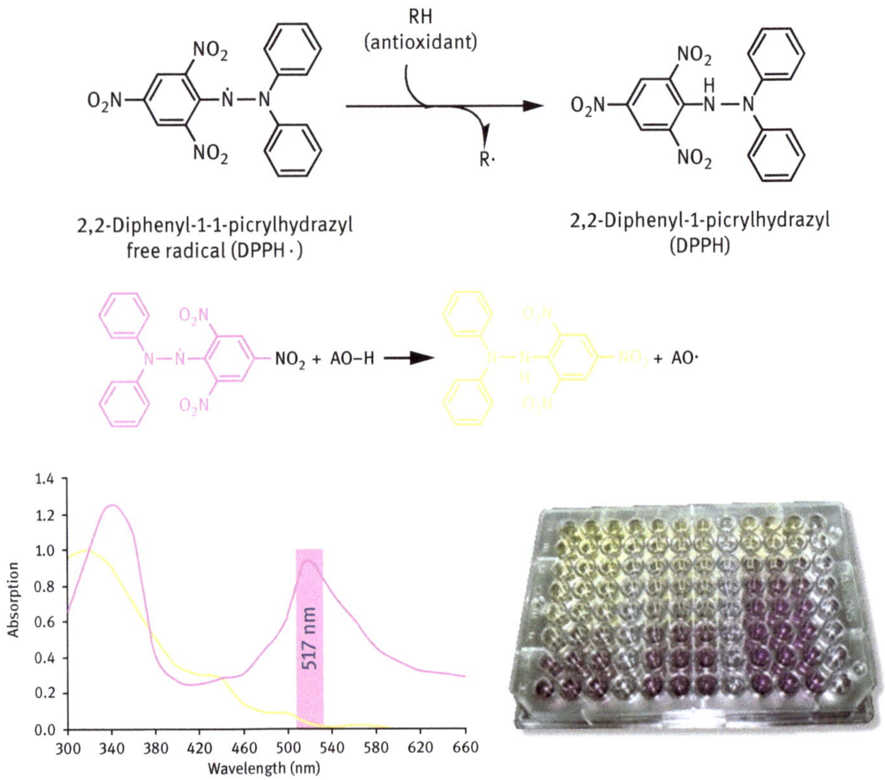

Figure 3.6: Determination of antioxidant capacity by DPPH spectroscopy.

3.2.4 ABTS reagent decolorization analysis

ABTS (2,2′-diazobis (3-ethylbenzothiazole-6-sulfonic acid)-diammonium salt) is a kind of cationic radicals, which have special optical properties (absorption peak positions: 645, 734, and 815 nm). It can be employed for nonbioactive antioxidant analysis. The decoloration assay measures the antioxidant capacity of antioxidant mixtures (Figure 3.7) and distinguishes synergistic effects between the components [46–49]. The method is adapted to sample analysis in a wide range of pH values [50]. The determination of samples in the pH range [50] can also be used to determine the antioxidant capacity of water- and fat-soluble antioxidants [51–52]. The major disadvantage of this method is that ABTS+ has a poor selectivity for hydrogen-donor reactions and can react with any hydroxylated aromatics that do not contribute to the antioxidant properties, making the process time-consuming.

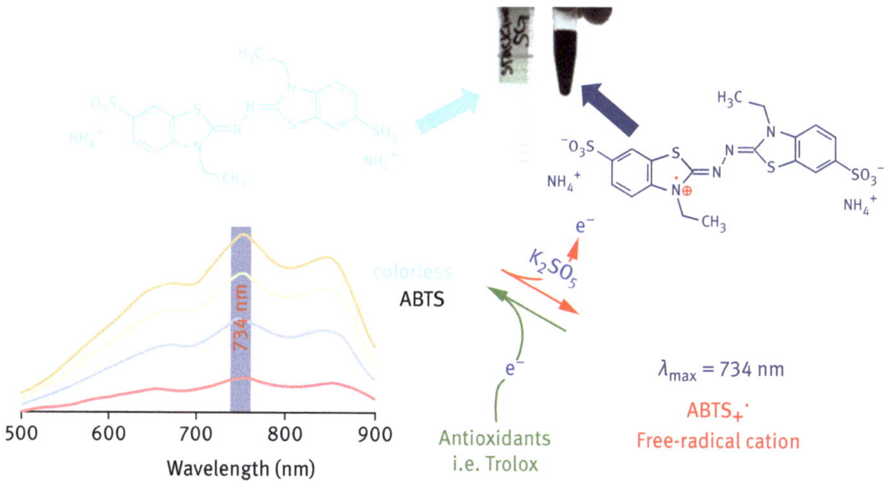

Figure 3.7: Determination of antioxidant capacity by TEAC (Trolox Equivalent Antioxidant Capacity) spectroscopy.

3.2.5 Determination of antioxidant capacity by Folin–Ciocalteu (F–C) method

$$Na_2WO_4/Na_2MoO_4 \rightarrow (\text{Phenol} - MoW_{11}O_{40})^{4-}$$

The F–C method is mainly applied for the determination of the total phenolic content, the change of absorbance between a wavelength of 745–750 nm [53–56], and then the antioxidant capacity of the phenolic antioxidant. The reagents used in the F–C method are simple, sensitive, and inexpensive; however, they are still affected by sugars and amines. Some inorganic ions, such as magnesium sulfate, ferric sulfate, and potassium nitrate, also affect the results.

The optical method has the following advantages: it is simple and cost effective, reagent strips can be easily prepared, and it can conveniently detect antioxidant capacity. However, the method suffers the drawback of being time-consuming as well as the interference of colored matters with the spectroscopic methods.

References

[1] Prior R L, Wu X, Schaich K. Standardized methods for the determination of antioxidant capacity and phenolics in foods and dietary supplements. Journal of Agricultural and Food Chemistry. 2005, 53: 4290–4302

[2] Davalos A, Gomez-Cordoves C, Bartolome B. Extending applicability of the oxygen radical absorbance capacity (ORAC–Fluorescein) assay. Journal of Agricultural and Food Chemistry, 2004, 52: 48–54.

[3] Ehlenfeldt M K, Prior R L. Oxygen radical absorbance capacity (ORAC) and phenolic and anthocyanin concentrations in fruit and leaf tissues of highbush blueberry. Journal of Agricultural and Food Chemistry, 2001, 49: 2222–2227.

[4] Prior R L, Hoang H, Gu L W, et al. Assays for hydrophilic and lipophilic antioxidant capacity (oxygen radical absorbance capacity (ORACFL)) of plasma and other biological and food samples. Journal of Agricultural and Food Chemistry, 2003, 51: 3273–3279.

[5] Yalcin G, Sogut O. Influence of extraction solvent on antioxidant capacity value of oleaster measured by ORAC method. Natural Product Research, 2014, 28: 1513–1517.

[6] Wayner D D M, Burton G W, Ingold K U, et al. Quantitative measurement of the total, peroxyl radical-trapping antioxidant capability of human blood plasma by controlled peroxidation: The important contribution made by plasma proteins. Febs Letters, 1985, 187: 33–37.

[7] Ghiselli A, Serafini M, Natella F, et al. Total antioxidant capacity as a tool to assess redox status: Critical view and experimental data. Free Radical Biology and Medicine, 2000, 29: 1106–1114.

[8] Ghiselli A, Serafini M, Maiani G, et al. A fluorescence method for measuring total plasma antioxidant capability. Free Radical Biology and Medicine, 1995, 18: 29–36.

[9] Cian R E, Salgado P R, Drago S R, et al. Development of naturally activated edible films with antioxidant properties prepared from red seaweed Porphyra columbina biopolymers. Food Chemistry, 2014, 146: 6–14.

[10] Popov I, Lewin G. Photochemiluminescent detection of antiradical activity. VI. Antioxidant characteristics of human blood plasma, low density lipoprotein, serum albumin and amino acids during in vitro oxidation. Luminescence, 1999, 14: 169–174.

[11] Popov I, Lewin G. Photochemiluminescent detection of antiradical activity. VII. Comparison with a modified method of thermo-initiated free radical generation with chemiluminescent detection. Luminescence, 2005, 20: 321–325.

[12] Popov I, Volker H, Lewin G. Photochemiluminescent detection of antiradical activity. V. Application in combination with the hydrogen peroxide-initiated chemiluminescence of blood plasma proteins to evaluate antioxidant homeostasis in humans. Redox Report, 2001, 6: 43–48.

[13] Mikami I, Yamaguchi M, Shinmoto H, et al. Development and validation of a microplate-based beta-carotene bleaching assay and comparison of antioxidant activity (AOA) in several crops

measured by beta-carotene bleaching, DPPH and ORAC assays. Food Science and Technology Research, 2009, 15: 171–178.
[14] Prieto Lage M A, Murado Garcia M A, Vazquez Alvarez J A, et al. A new microplate procedure for simultaneous assessment of lipophilic and hydrophilic antioxidants and pro-oxidants, using crocin and β-carotene bleaching methods in a single combined assay: Tea extracts as a case study. Food Research International, 2013, 53: 836–846.
[15] Burda S, Oleszek W. Antioxidant and antiradical activities of flavonoids. Journal of Agricultural and Food Chemistry, 2001, 49: 2774–2779.
[16] Jorgensen L V, Andersen H J, Skibsted L H. Kinetics of reduction of hypervalent iron in myoglobin by crocin in aqueous solution. Free Radical Research, 1997, 27: 73–87.
[17] Winston G W, Regoli F, Dugas A J, et al. A rapid gas chromatographic assay for determining oxyradical scavenging capacity of antioxidants and biological fluids. Free Radical Biology and Medicine, 1998, 24: 480–493.
[18] Regoli F, Winston G W. Quantification of total oxidant scavenging capacity of antioxidants for peroxynitrite, peroxyl radicals, and hydroxyl radicals. Toxicology and Applied Pharmacology, 1999, 15: 96–105.
[19] Koren E, Kohen R, Ginsburg I. Polyphenols enhance total oxidant scavenging capacities of human blood by binding to RBC. Experimental Biology and Medicine, 2010, 235: 689–699.
[20] Ryu C S, Kim C H, Lee S Y, et al. Evaluation of the total oxidant scavenging capacity of saponins isolated from Platycodon grandiflorum. Food Chemistry, 2012, 132: 333–337.
[21] Alho H, Leinonen J. Total antioxidant activity measured by chemiluminescence methods. Oxidants and Antioxidants, 1999, 299: 3–15.
[22] Krol W, Czuba Z, Scheller S. Anti-oxidant property of ethanolic extract of propolis (EEP) as evaluated by inhibiting the chemiluminescence oxidation of luminol. Biochemistry International, 1990, 21: 593–597.
[23] Lissi E, Salimhanna M, Pascual C. Evaluation of total antioxidant potential (TRAP) and total antioxidant reactivity from luminol-enhanced chemiluminescence measurements. Free Radical Biology and Medicine, 1995, 18: 153–158.
[24] Malejko J, Nalewajko-Sieliwoniuk E, Nazaruk J, et al. Determination of the total polyphenolic content in Cirsium palustre (L.) leaves extracts with manganese(IV) chemiluminescence detection. Food Chemistry, 2014, 152: 155–161.
[25] Zargoosh K, Ghayeb Y, Aeineh N, et al. Evaluation of antioxidant capacity of hydrophilic and hydrophobic antioxidants using peroxyoxalate chemiluminescence reaction of the novel furandicarboxylate derivative. Food Analytical Methods, 2014, 7: 283–290.
[26] Handelman G J, Cao G, Walter M F, et al. Antioxidant capacity of oat (Avena sativa L.) extracts. I. Inhibition of low-density liporotein oxidation and oxygen radical absorbance capacity. Journal of Agricultural and Food Chemistry, 1999, 47: 4888–4893.
[27] Neudorffer A, Bonnefont-Rousselot D, Legrand A, et al. 4-hydroxycinnamic ethyl ester derivatives and related dehydrodimers: Relationship between oxidation potential and protective effects against oxidation of low-density lipoproteins. Journal of Agricultural and Food Chemistry, 2004, 52: 2084–2091.
[28] Filipe P, Haigle J, Freitas J, et al. Anti- and pro-oxidant effects of urate in copper-induced low-density lipoprotein oxidation. European Journal of Biochemistry, 2002, 269: 5474–5483.
[29] Sahebkar A, Emami S A, Asgary S, et al. Antioxidant activity of essential oils from cupressus species against oxidation of LDL. Circulation, 2012, 125: E912–E913.
[30] Benzie I F F, Strain J J. The ferric reducing ability of plasma (FRAP) as a measure of "Antioxidant Power": The FRAP assay. Analytical Biochemistry, 1996, 239: 70–76.

[31] Hadipour H A, Daryadel A, Allameh A. Comparative effects of vitamin K1 on the total antioxidant capacity in adult and weanling rats using 'ferric reducing ability of plasma'. Toxicology Letters (Shannon), 2001, 123: 33.
[32] Amamcharla J K, Metzger L E. Modification of the ferric reducing antioxidant power (FRAP) assay to determine the susceptibility of raw milk to oxidation. International Dairy Journal, 2014, 34: 177–179.
[33] Biskup I, Golonka I, Gamian A, et al. Antioxidant activity of selected phenols estimated by ABTS and FRAP methods. Postepy Higieny I Medycyny Doswiadczalnej, 2013, 67: 958–963.
[34] Martins A C, Bukman L, Vargas A M M, et al. The antioxidant activity of teas measured by the FRAP method adapted to the FIA system: Optimizing the conditions using the response surface methodology. Food Chemistry, 2013, 138: 574–580.
[35] Apak R, Guclu K, Ozyurek M, et al. Novel total antioxidant capacity index for dietary polyphenols and vitamins C and E, using their cupricÎon reducing capability in the presence of neocuproine: CUPRAC method. Journal of Agricultural and Food Chemistry, 2004, 52: 7970–7981.
[36] Celik S E, Ozyurek M, Guclu K, et al. Differences in responsivity of original cupric reducing antioxidant capacity and cupric–bathocuproine sulfonate assays to antioxidant compounds. Analytical Biochemistry, 2012, 423: 36–38.
[37] Ozyurek M, Guclu K, Apak R. The main and modified CUPRAC methods of antioxidant measurement. Trac-Trends in Analytical Chemistry, 2011, 30: 652–664.
[38] Bekdeser B, Ozyurek M, Guclu K, et al. Development of a new catalase activity assay for biological samples using optical cuprac sensor. Spectrochimica Acta Part a-Molecular and Biomolecular Spectroscopy, 2014, 132: 485–490.
[39] Bener M, Ozyurek M, Guclu K, et al. Novel optical fiber reflectometric cuprac sensor for total antioxidant capacity measurement of food extracts and biological samples. Journal of Agricultural and Food Chemistry, 2013, 61: 8381–8388.
[40] Brandwilliams W, Cuvelier M E, Berset C. Use of free radical method to evaluate antioxidant activity. Food Science and Technology-Lebensmittel- Wissenschaft and Technologie, 1995, 28:25–30.
[41] Sanchez-Moreno C, Larrauri J A, Saura-Calixto F. A procedure to measure the antiradical efficiency of polyphenols. Journal of the Science of Food and Agriculture, 1998, 76: 270–276.
[42] Kim M H, Jang J H, Oh M H, et al. The comparison of DPPH-scavenging capacity and anti-inflammatory effects of phenolic compounds isolated from the stems of Stewartia koreana Nakai. Natural Product Research, 2014, 28: 1409–1412.
[43] Tan C, Xue J, Abbas S, et al. Liposome as a delivery system for carotenoids: Comparative antioxidant activity of carotenoids as measured by ferric reducing antioxidant power, DPPH assay and lipid peroxidation. Journal of Agricultural and Food Chemistry, 2014, 62: 6726–6735.
[44] Xie J, Schaich K M. Re-evaluation of the 2,2-Diphenyl-1-picrylhydrazyl free radical (DPPH) assay for antioxidant activity. Journal of Agricultural and Food Chemistry, 2014, 62: 4251–4260.
[45] Mishra K, Ojha H, Chaudhury N K. Estimation of antiradical properties of antioxidants using DPPH• t assay: A critical review and results. Food Chemistry, 2012, 130: 1036–1043.
[46] Miller N J, Riceevans C, Davies M J, et al. A novel method for measuring antioxidant capacity and its application to monitoring the antioxidant status in premature neonates. Clinical Science, 1993, 84: 407–412.
[47] Berg R V D, Haenen G, Berg H V D, et al. Applicability of an improved Trolox equivalent antioxidant capacity (TEAC) assay for evaluation of antioxidant capacity measurements of mixtures. Food Chemistry, 1999, 66: 511–517.

[48] Magalhaes L M, Barreiros L, Reis S, et al. Kinetic matching approach applied to ABTS assay for high-throughput determination of total antioxidant capacity of food products. Journal of Food Composition and Analysis, 2014, 33: 187–194.
[49] Tian X, Schaich K M. Effects of molecular structure on kinetics and dynamics of the trolox equivalent antioxidant capacity assay with ABTS+•. Journal of Agricultural and Food Chemistry. 2013, 61: 5511–5519.
[50] Cano A, Hernandez-Ruiz J, Garcia-Canovas F, et al. An end-point method for estimation of the total antioxidant activity in plant material. Phytochemical Analysis, 1998, 9: 196–202.
[51] Cano A, Acosta M, Arnao M B. A method to measure antioxidant activity in organic media: Application to lipophilic vitamins. Redox Report, 2000, 5: 365–370.
[52] Cano A, Alcaraz O, Acosta M, et al. On-line antioxidant activity determination: Comparison of hydrophilic and lipophilic antioxidant activity using the ABTS•+ assay. Redox Report, 2002, 7: 103–109.
[53] Ainsworth E A, Gillespie K M. Estimation of total phenolic content and other oxidation substrates in plant tissues using Folin-Ciocalteu reagent. Nature Protocols, 2007, 2: 875–877.
[54] Berker K I, Olgun F A O, Ozyurt D, et al. Modified Folin-Ciocalteu antioxidant capacity assay for measuring lipophilic antioxidants. Journal of Agricultural and Food Chemistry, 2013, 61: 4783–4791.
[55] Everette J D, Bryant Q M, Green A M, et al. Thorough study of reactivity of various compound classes toward the Folin-Ciocalteu reagent. Journal of Agricultural and Food Chemistry, 2010, 58: 8139–8144.
[56] Medina-Remon A, Barrionuevo-Gonzalez A, Zamora-Ros R, et al. Rapid Folin-Ciocalteu method using microtiter 96-well plate cartridges for solid phase extraction to assess urinary total phenolic compounds, as a biomarker of total polyphenols intake. Analytica Chimica Acta, 2009, 634: 54–60.

4 Determination of antioxidant capacity of chromatography

Chromatographic separation invented by the Russian botanist Mikhail Tswett in 1903 has now become one of the most important analytical techniques, and has been widely used in petrochemical, organic synthesis, medicine and health, environmental science and many other research areas.

Chromatography (Figure 4.1) utilizes the selective partitioning of different substances in different phases for the elution of the mixture in the mobile phase with respect to the stationary phase. Different substances in the mixture move along the stationary phase at different velocities, thus achieving a separation effect. Because of its advantages of having high separation efficiency, fast analysis, high detection sensitivity, small sample amount, simultaneous analysis of multiple components and easy automation, the chromatographic method has become the most widely used and rapidly developed technique in analytical chemistry.

In the late 1960s, high-performance liquid chromatography (HPLC) was developed for technical requirements for the separation and analysis of complex biological macromolecules such as proteins and nucleic acids. At present, HPLC developed in the field of organic, biological, pharmaceutical, chemical, environmental and other fields, especially in the field of food science it has been widely used, and has become the most commonly used separation and detection tool.

Chromatography separates antioxidants individually in complex food components and differentiates them [1–9]. To improve the accuracy and efficiency of the separation, mass spectrometry method is used for abundant structural information, and the technical methods of HPLC and mass spectrometry are introduced into the field of antioxidant detection and analysis [10–17].

The results obtained from chromatography are the concentration of each antioxidant. Although the concentration of antioxidant can reflect the antioxidant capacity to a certain extent, the antioxidant capacity of different antioxidants cannot be compared with the standard. To determine the antioxidant capacity, electrochemical detectors were introduced for analysis [18–22], but even with the introduction of an electrochemical detector, the synergy between the antioxidants cannot be effectively measured by chromatography, especially HPLC. Another disadvantage of chromatography coupled with mass spectrometry is that the instrument system is expensive and complex to operate, making it difficult to be used universally in common laboratories.

https://doi.org/10.1515/9783110573763-004

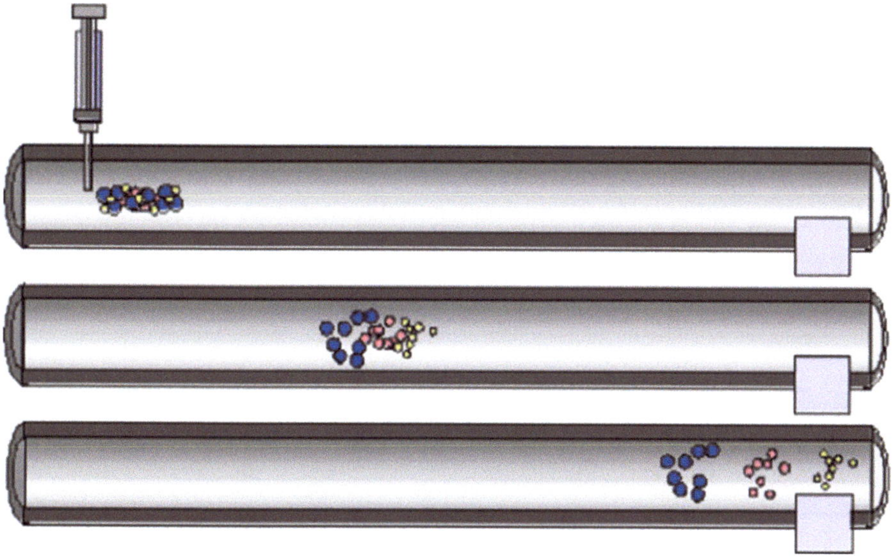

Figure 4.1: Schematic diagram of chromatographic analysis techniques.

4.1 Enrichment and separation of an antioxidant sample

4.1.1 Extraction methods

1 Traditional extraction methods

(1) Solvent extraction method: The method where different samples of different components in various solvents with varied solubilities and similar principles of compatibility extract samples of the antioxidant component is known as solvent extraction. Solvent extraction is the most common antioxidant extraction method, which is mainly divided into dipping method, percolation method, boiling method and reflux method. At the time of extraction, a suitable solvent should be selected according to the polarity and water solubility of the extracted components. For example, when a strong polar flavone is extracted, a mixed solvent such as methanol–water and ethanol–water is usually used; while a weak polar flavone is extracted, solvents such as ethyl acetate, ether, and chloroform are usually applied. The most commonly used solvents are water and ethanol. Methanol extraction works well, but because of its toxicity, ethanol is generally used instead. Zhang Haiying et al. [23] studied that in the extraction of antioxidant components of hawthorn, solvent extraction method was used to compare the reflux ratio of different solid–liquid ratio, extraction time, the number of extraction and ethanol concentration in the extraction of flavonoids after repeated tests to determine the

best extraction conditions for 50% ethanol, 1:14 liquid ratio, extracted for 2 times each at 1.5 h.

(2) Decompression sublimation method: This method is mainly applied for sublimation in the extraction of active ingredient. For example, Tao Yong Yuan et al. [24] used vacuum distillation method to extract tea polyphenols, pressure 8 × 104 Pa, temperature 145 °C, time 25 min, which were the best extraction conditions. The sublimation extraction of high purity, low cost, no chemical residues is shown in Figure 4.2.

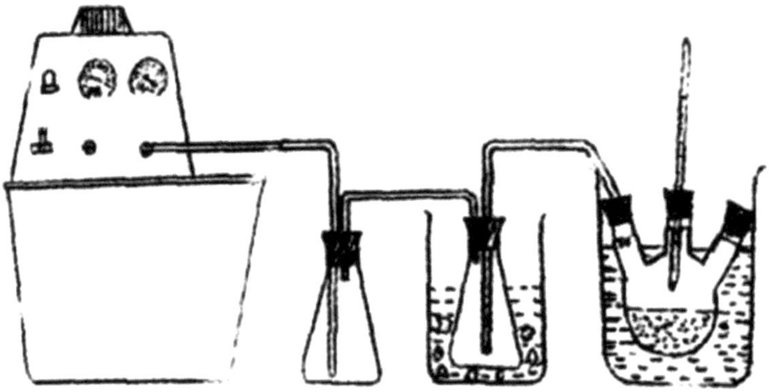

Figure 4.2: Scheme of tea polyphenol extraction device in dehydration sublimation.

(3) Steam distillation: When the extracted material is insoluble in water and does not react with water, you can use steam distillation.

(4) Squeezing method: When the samples of the antioxidant component are high, you can use the press method to extract fresh samples of antioxidant components.

2 Modern extraction of new technologies

(1) Ultrasonic extraction: Ultrasonic extraction is the use of ultrasonic radiation pressure to produce mechanical stirring, where solute diffusion occurs for accelerating a new extraction technology. For example, Wang Yanfeng et al. [25] extracted the flavonoids from leaves of *Ginkgo biloba L* by ultrasonic method. Compared with the continuous refluxing Soxhlet extraction method, this method has obvious advantages. The optimum operating conditions are: ultrasonic frequency 40k Hz, treatment time 10 min and resting time 12 h (Figure 4.3).

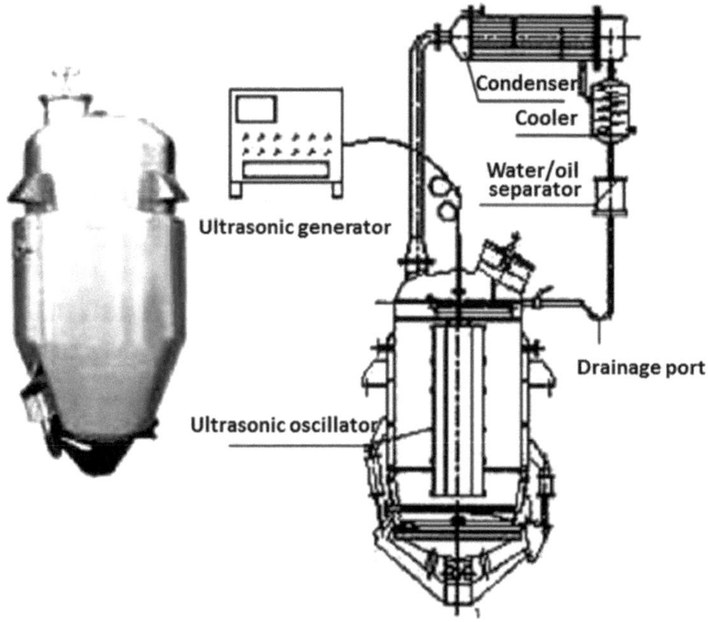

Figure 4.3: A typical schematic device of ultrasonic extraction.

(2) Microwave extraction: Microwave-assisted extraction technology is the use of microwave heat generated by the dissolution of solute to accelerate diffusion of a new extraction technology. This technique is widely used because of its simplicity and speed, low solvent demand and environmental friendliness. For example, Guo Zijie et al. [26] studied the microwave extraction of ginseng saponins in *Panax notoginseng* that has the best extraction conditions of ultrasonic power of 600W, 80% methanol, solid–liquid ratio of 1:4; Liu Yuan et al. [27] studied the extraction of antioxidant components in Gold lotus using the microwave extraction method with the highest extraction efficiency with 70% ethanol as a solvent at 75 °C, low fire microwave at 5 min, ratio of 1:20. Because of its short time, simple operation and high extraction rate, the extraction technique has been widely applied in the fields of chemistry, biology and medicine.

(3) Supercritical fluid extraction technology: The supercritical fluid extraction (SFE) technology is a solvent that has dual properties of gas and liquid at critical point. Utilizing the fluid near the critical state as the extraction solvent, pressure conditions are applied for extracting different polarities of the sample composition. This technique is suitable for extracting lipophilic ingredients from samples (Figure 4.4).

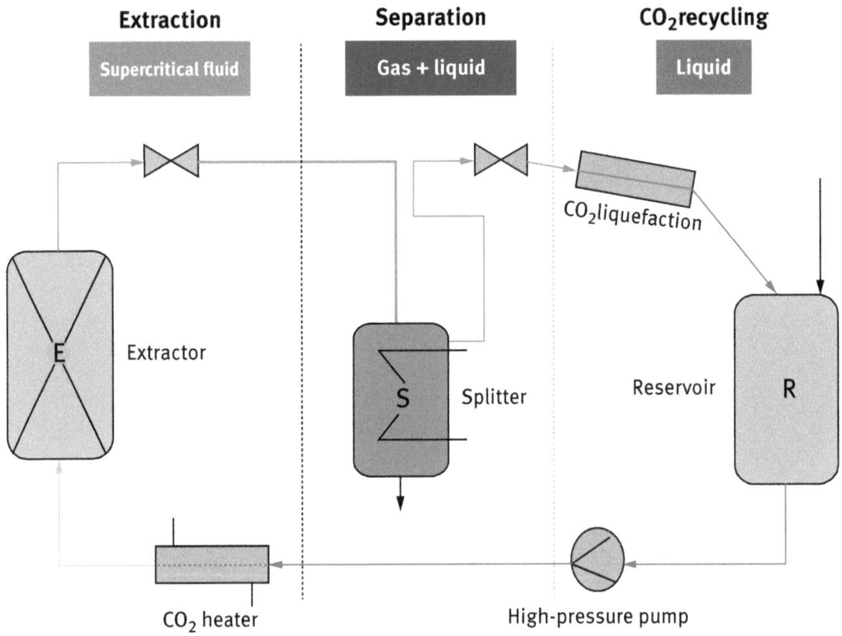

Figure 4.4: Supercritical CO_2 fluid extraction and extraction device.

Carbon dioxide, nitrogen dioxide, water, ethane and similar gases are commonly used as supercritical fluids. Because of its colorless, odorless and nontoxic nature, carbon dioxide is stable in nature and low in cost. The critical temperature is low (31.3 °C) and the critical pressure is 7.2 MPa. It is widely used in the SFE. However, when carbon dioxide is used as an extraction solvent, its polarity is less and it is only suitable for the extraction of small polar and nonpolar compounds. When extracting compounds with large polarities, an entrainer is be added to increase the solubility of the extracted components. An entrainer is a polar solvent, commonly used in methanol, ethanol and so on, and can improve the extraction efficiency.

The main influencing factors of SFE technology are extraction temperature, extraction time, extraction pressure, entrainer and so on. For example, Jinrucheng et al. [28] used the supercritical carbon dioxide extraction technique to extract isoflavone from *Trifolium pratense L.* The optimum extraction conditions were as follows: temperature 45 °C, pressure of 25 MPa with ethanol as an entrainer; the flow of carbon dioxide is 6 mL/min, 40 mesh raw material, static extraction of 60 min and dynamic extraction at 15 min.

(4) Pressurized solvent extraction: Under high pressure, the diffusion speed of solute is accelerated; the solvent is quickly immersed in the pores of the extract, the concentration of extract solution and the external solution is achieved to

maintain balance in a short period of time. As the pressure is gradually reduced, the pressure of the external solvent drastically decreases and the cells rupture. The rapid transfer of intracellular substances into the solvent increases the extraction rate. This method has the advantage of short time, high efficiency, less energy consumption, simple operation, high extraction rate, green environmental protection, low extraction temperature and can be carried out at normal temperature to avoid the damage to the extracted substances caused by overheat, and this is a novel extraction method. At present, this method has been widely used in many fields such as food, biochemistry, pharmacy and so on, such as food quality improvement, food safety control and extraction of active ingredients in traditional Chinese medicine.

4.1.2 Separation and purification methods

1 Traditional separation and purification methods

(1) Extraction method: This method is for extraction of substances in the partition coefficient of an immiscible solvent to separate substances.
(2) Precipitation method: This method includes a precipitating agent into the extracting solution to precipitate the target component or the impurity component.

For separation:

(3) Crystallization method: According to different solubilities of different substances in the extract, the crystallization method is used for achieving the purpose of separation.
(4) Salting-out method: This method includes the addition of certain amount of salt in the extracting solution to reduce the solubility of the substance to be separated.

For precipitation:

(5) Dialysis method: This method demonstrates that in solution, small molecules can pass through a semipermeable membrane, and macromolecules cannot be semipermeable. membranes. According to this principle, the material extracted after the separation method is called the dialysis method.
(6) Column chromatography: Column chromatography is based on the extraction of different components in the stationary phase and flow.

The partition coefficient in the phase varies after adsorption and desorption to achieve the purpose of separation. Column chromatography is mainly based on silica.

There are five types of chromatography: (1) gel, (2) aluminum oxide column chromatography, (3) polyamide column chromatography, (4) macroporous adsorption resin column chromatography and (5) dextrose gel column chromatography.

2 Modern separation and purification of new technologies

(1) Foam separation technology: The foam separation technology is a new type of separation and purification technology developed in the last 10 years. This is based on the principle that for a bubble in a liquid containing surfactant, the active material is gathered on the surface of the bubble, so that the liquid phase below has been purified, which helps in separation. It is noteworthy that the material in the foam separation technology must be separated from the surface-active substances. Gums in Chinese herbal medicine, saponins and so on have surface-active ingredients, so that the separation and purification can choose foam separation technology. The equipment for foam separation technology is simple, easy to operate and environmental pollution is also less; however, as of its separation is selective, and the separation of the solution is based on pH, temperature, concentration, gas flow velocity, ionic strength, stirring and other factors, its application has been limited.

(2) Membrane separation technology: Membrane separation separates solute and solvent based on the membrane. As the solution passes through the membrane, solvent molecules and some small molecules can pass through the membrane, whereas solutes and some large molecules are trapped in the membrane. Membrane separation techniques include electrodialysis, diffusion dialysis, ultrafiltration and reverse osmosis. Flavonoids are mostly of the molecular weight of 100 or less, and polysaccharides, macromolecules have molecular weight more than 50,000, so the membrane separation technology can effectively separate flavonoids and other macromolecules to achieve the purpose of separation. This separation technology is simple, easy to operate, has low energy consumption, high extraction rate, and does not require heating.

(3) High-speed countercurrent chromatography: The principle of high-speed countercurrent chromatography (HSCCC) is based on the fact that the two-phase solvent is effectively mixed and distributed under the effect of asymmetric centrifugal force for the separation of substances in the two-phase solvent. In the study of Chinese herbal antioxidant, high-efficiency countercurrent chromatography has a wide range of applications. For example, Cai Dingguo et al. [29] studied that HSCC flavonoids were extracted from antioxidant components of *Ginkgo biloba*.

4.2 High-performance liquid chromatography for the determination of antioxidants

Chromatography is based on the selective distribution of different substances in different phases, which can separate the antioxidants in food one-by-one for detection.

Rong Tsao et al. used HPLC to investigate the in-depth differences in antioxidative components caused by different varieties and locations of apples planted around the Lake Ontario [8]. In general, polyphenol content in apple peel is much better than that in apple flesh. The differences in the varieties significantly affect the total polyphenol content in fruits. Taking the examples of the eight varieties of apples studied, the five major classes of antioxidants commonly found in apples are as follows: (1) cinnamic acid, (2) flavan-3-ol/proanthocyanidins, (3) anthocyanins, (4) flavonols and (5) ketones; 16 antioxidants have been identified (Figure 4.5). Hydroxycinnamate, delphinidin and flavan-3-ol are found extensively in the peel and pulp, whereas quercetin appears almost exclusively in the peel. The antioxidant cyanidin galactoside is found only in red apple peel; apple peel and pulp content are the largest proanthocyanidins, followed by quercetin glycosides or dihydroxycinnamate; 3-hydroxy phloretin-2x oligosaccharide glycosides was first detected in apples. These findings are conducive to the understanding of the type and content of apple polyphenols, and provide a clear technical guideline for healthy nutrition intake.

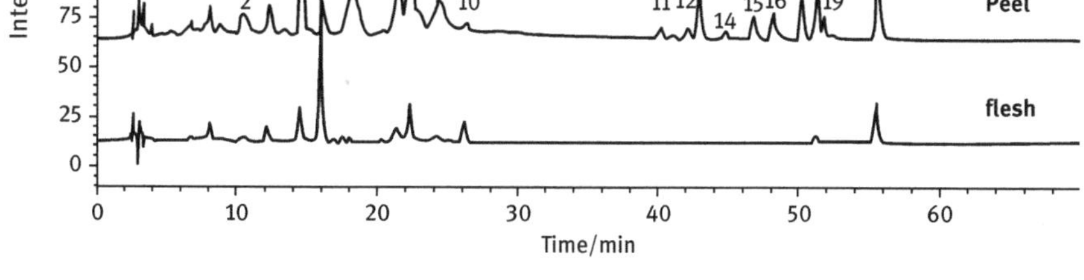

Figure 4.5: Chromatogram of red snake fruit peel and pulp extract at 280 nm. 1 – proanthocyanidin B1; 2 – unknown; 3 – catechin; 4 – proanthocyanidin B2; 5 – chlorogenic acid; 6 – cyanidin galactoside; 7 – unknown proanthocyanidin dimer; 8 – epicate catechin; 9 – unknown proanthocyanidin dimer; 10 – cinnamoyl quinic acid; 11 – unknown proanthocyanidin dimer; 12 – 3-hydroxyphloretin-2′-xyloglucoside; 13 – Quercetin-3-galactoside; 14 – quercetin-3-glucoside; 15 – quercetin-3-xyloside; 16 – 3-hydroxy-peptidin-2′-glucoside; 17 – skin 3-arabinoside; 18 – root dermatin-2′-xylose glucoside; 19 – quercetin-3-rhamnoside; 20 – root glucoside.

Fukumoto et al. [5] designed a HPLC method for the quantitative analysis of malondialdehyde produced in the Cu^{2+}-catalyzed linoleic acid emulsion oxidation system to evaluate the pro-oxidant and antioxidant activities of the phenolic compounds. Studies have shown that most standard antioxidants, such as standard phenolic acid, flavonol, flavanol, flavanone and anthocyanin, or the actual phenolic compounds isolated from berry extracts, have the potential to be most effective at low concentration oxidation (Tables 4.1 and Table 4.2). Antioxidant activity is at maximum when the concentration is increased to a certain level.

The research further compares HPLC with the results of β-carotene bleaching method and DPPH method. The results showed that although the reaction conditions and mechanism of the three methods are different, the measured results are not the same; however, for the structurally similar compounds, the observed trend of its antioxidant activity is consistent. In general, the antioxidant activity increases with the increase of hydroxyl groups contained in the molecular structure, while the antioxidant activity decreases with the increase in glycosylation groups. In terms of antioxidant properties, both HPLC and β-carotene bleaching methods demonstrate that naturally occurring antioxidants are in no way inferior to chemically synthesized antioxidants, providing a reliable screening route for food processing companies to choose anti-rancid additives.

4.3 Instrumentation technology for the determination of antioxidants

To improve the accuracy and efficiency of chromatographic separation, technical methods coupled with mass spectrometry and similar methods have also been introduced into the field of antioxidant analysis.

Chandra et al. [10] established HPLC and LC/ES–MS methods to quantitatively analyze the contents of anthocyanins and their total contents in herbal raw materials. This simple and easy method eliminates the need for any solid-phase extraction (SPE) sample processing. Using cyanidin-3-glucoside chloride as the external standard equivalent and multiplying by the molecular weight correction factor gives you the ability to quantify the total anthocyanin content. This mode is also suitable for routine quality control of anthocyanin-containing raw materials. This method can effectively provide a more accurate technical reference for the field of nutritional supplements.

Ayaz et al. [12] successfully applied HPLC–MS technology for qualitative and quantitative determination of phenolic acids in blueberries. It has been confirmed that berry contains gallic acid, protocatechuic acid, *p*-hydroxybenzoic acid, *m*-hydroxybenzoic acid, gentisic acid, chlorogenic acid, *p*-coumaric acid, caffeic acid, ferulic acid, erucic acid, salicylic acid and trans-cinnamic acid (Table 4.3), indicating that fruit can be considered as a potentially important source of phenolic acids.

Table 4.1: Oxidative and antioxidant activities of typical polyphenolic compounds.

Compound	Current study				Selected results from the literature	
	β-Carotene method	DPPH* method	HPLC method		DPPH* assay[a]	ORAC assay[b]
	Initial slope[c] ($\times 10^{-6}$)	Antiradical activity[c]	Antioxidant activity[e] (μM) of compound added	Prooxidant activity[f]	Antiradical powers[g] (antiradical activity)	ORAC slope[h]
Phenolic adds						
Benzoic acid derivatives						
Gallic acid	630 ± 38	−6.21 ± 0.60	1,500–2,000	+	12.5 [−6.25]	
Protocatechuic acid	nc[i]	−4.56 ± 0.09	3,000–3,500	+	7.14 [−3.57]	
3-Hydroxybenzoic acid	nc	nc	>4,000	+		
Vanillic acid	nc	−0.99 ± 0.31	>4,000	+	0.17 [−0.09]	
Syringic acid	nc	−3.03 ± 0.07	>4,000	+		
Ellagic acid	779 ± 83	−9.21 ± 0.25	>4,000	+		
Cinnamic acid derivatives						
Caffeic acid	620 ± 36	−4.49 ± 0.24[j]	500–1,000	+	9.1 [−4.55]	
p-Coumaric acid	nc	−0.33 ± 0.06	>4,000	+	0.02 [−0.01]	

(continued)

Table 4.1 (continued)

Compound	**Current study**				**Selected results from the literature**	
	β-Carotene method	**DPPH* method**	**HPLC method**		**DPPH* assay**[a]	**ORAC assay**[b]
	Initial slope[c] **($\times 10^{-6}$)**	**Antiradical activity**[c]	**Antioxidant activity**[e] **(μM) of compound added**	**Prooxidant activity**[f]	**Antiradical powers**[g] **(antiradical activity)**	**ORAC slope**[h]
Cinnamic acid	nc	nc	>4,000	nd[k]		
Ferulic acid	161 ± 14	−1.34 ± 0.05	>4,000	+	2.33 [−1.17]	
Chlorogenic acid	186 ± 26	−5.08 ± 0.29	1,000–1,500	+		
			Flavonols			
Myricetin	1,046 ± 83	−6.59 ± 0.57	500–1,000	+		4.319 ± 0.119
Quercetin	630 ± 42	−6.73 ± 0.08	200–300	+		3.285 ± 0.117
Rutin	nc	−5.10 ± 0.10	500–1,000	+		
Kaempferol	172 ± 6	−2.09 ± 0.10	>4,000	+		2.671 ± 0.131
			Flavanols			
(+)-Catechin	443 ± 63	−7.19 ± 0.32	500–1,000	+		
(−)-Epicatechin	515 ± 39	−7.65 ± 0.17	500–1,000	+		

			Flavanones			
Naringenin	nc	−0.18 ± 0.01	>4,000	+		
			Anthocyanidins/anthocyanins			
Cyanidin	836 ± 69	−7.40 ± 0.15	200–300	+		2.239 ± 0.029
Cyanidin 3-glucoside	278 ± 32	−6.81 ± 0.30	300–400	+		3.491 ± 0.011
Cyanidin 3,5-diglucoside	220 ± 39	−3.32 ± 0.07	500–1,000	+		1.689 ± 0.052
Delphinidin	897 ± 147	−8.86 ± 0.28	500–1,000	+		1.809 ± 0.068
Malvidin	288 ± 34	−4.49 ± 0.28	1,500–2,000	+		2.009 ± 0.167
Malvidin 3-glucoside	448 ± 40	−4.29 ± 0.42	500–1,000	+		1.404 ± 0.052
Malvidin 3,5-diglucoside	266 ± 27	−2.56 ± 0.10	2,000–2,500			−1.550 ± 0.062
Pelargonidin	nc	−4 63 ± 0.25	1,500–2,000	+		1.540 ± 0.033
Pelargonidin 3-glucoside	444 ± 94	−3.95 ± 0.22	2,000–2,500	+		1.560 ± 0.145
Pelargonidin 3,5-diglucoside	nc	−2.04 ± 0.10	2,000–2,500	+		1.067 ± 0.043
Peonidin	169 ± 22	−4.05 ± 0.17	1,500–2,000	+		1.693 ± 0.035
Peonidin 3-glucoside	251 ± 4	−3.38 ± 0.15	2,500–3,000	+		1.805 ± 0.014
		Standards				
Ascorbic acid	nc	−1.83 ± 0.07	>4,000	+	3.7 [−1.85]	

(continued)

Table 4.1 (continued)

Compound	Current study				Selected results from the literature	
	β-Carotene method	DPPH* method	HPLC method		DPPH* assay[a]	ORAC assay[b]
	Initial slope[c] (× 10^{-6})	Antiradical activity[c]	Antioxidant activity[e] (μM) of compound added	Prooxidant activity[f]	Antiradical powers[g] (antiradical activity)	ORAC slope[h]
α-Tocopherol	870 ± 21	−1.95 ± 0.07	2,000–2,500 (50%)	nd		
BHA	835 ± 50	−2.61 ± 0.01	1,000–1,500	nd	4.17 [−2.09]	
BHT	664 ± 76	−3.17 ± 0.07	200–300	nd	4.2 [−2.1]	

[a] The DPPH* assay used by Brand-Williams et al. (1995).
[b] The oxygen radical absorbing capacity (ORAC) assay measures reaction of peroxyl radicals expressed as μM of Trolox equivalent per μM of compound. Results for flavonols were taken from the study of Cao et al. (1997), and results for anthocyanidins/anthocyanins were taken from Wang et al. (1997).
[c] Values are means of slope coefficients calculated by linear regression ± standard deviations (n = 3) in A_{450nm} that alter 90 min of incubation in the dark/μM of compound added.
[d] Values are means of slope coefficients calculated by linear regression ± standard deviations (n = 3) in μM of DPPH*/μM of compound.
[e] Antioxidant activity was defined by the concentration range of compound added to reach 0% malonaldehyde of the control.
[f] Prooxidant activity was positive (+) if the percentage of malonaldehyde of the control was >100% in the concentration range tested.
[g] Antiradical power was defined as the reciprocal of the amount of antioxidant needed to decrease the initial DPPH* concentration by 50%. The antiradical activity was equivalent to negative half of the antiradical power.
[h] Values are slope coefficients calculated by linear regression ± standard error.
[i] Not calculated since linear regression $r^2 < 0.800$.
[j] Values were obtained after reaction for 48 h.
[k] Not detected.

Table 4.2: Polyphenol content in berry extracts and their prooxidation/antioxidant activities.

Property	**Extract**			
	Blackberry	**Blackcurrant**	**Blueberry**	**Saskatoon berry**
	Phenolic content			
Total phenolics[a] (mg of chlorogenic acid/100 g)	383 ± 2	347 ± 3	393 ± 5	405 ± 2
Tartaric esters[a] (mg of caffeic acid/100 g)	27 ± 2	27 ± 2	55 ± 8	73 ± 2
Flavonols[a] (mg of quercetin/100 g)	24 ± 3	27 ± 3	46 ± 7	54 ± 1
Anthocyanins[a] (mg of malvidin 3-glucoside/100 g)	149 ± 2	213 ±3	233 ± 3	153 ± 2
	Antioxidant/prooxidant activity			
β-Carotene method[b] initial slope (× 10^{-6})	642 ± 10	885 ± 63	654 ± 72	575 ± 34
DPPH* method[c] antiradical activity	−5.27 ± 0.34	6.61 ± 0.96	3.64 ± 0.20	4.97 ±0.10
HPLC method antioxidant activity[d] (μM of total phenolic added)	401–680	489–979	223–446	230–460
HPLC method prooxidant activity[e]	+	+	+	+

[a] Values are means ± standard deviations (n = 3).
[b] Values are means of slope coefficients calculated by linear regression ± standard deviations (n = 3) in A_{450nm} after 90 min of incubation in the dark/μM of total phenolics as chlorogenic acid added.
[c] Values are means of slope coefficients calculated by linear regression ± standard deviations (n = 3) in μM of DPPH*/μM of total phenolics as chlorogenic acid.
[d] Antioxidant activity was defined by the concentration range of total phenolics as chlorogenic acid needed to reach 0% malonaldehyde of the control. Concentration range of total phenolics was the mean value (n = 3).
[e] The prooxidant activity was positive (+) if the percentage of malonaldehyde of the control was >100% in the concentration range tested.

For the quantitative analysis of catechin and gallic acid in green tea, Bedner et al. [16] proposed a method that nonlabeled internal standard substance hydroxypylline should be applied during the LC-ESI/MS detection. The corresponding dynamic calibration quantitative model originated from polynomial data fitting will effectively compensate in real time the changes of response factor (Figures 4.6 and 4.7). Although the dynamic correction of MS data is a difficult and time-consuming task, the quantitative data obtained by this method is very reliable compared to the UV detection that would be affected by interferences.

Table 4.3: Phenolic acid content (ng/g) extracted from blueberries in northeastern Anatolia.

Phenolic acid	Phenolic fraction				
	Free	Esters	Glycoside	Ester-bound	Total[b]
Gallic acid	2.8 ± 0.3	3.0 ± 0.2	ND[c]	73.0 ± 8.1	78.8
Protocatechuic acid	56.1 ± 10.8	451.0 ± 14.8	481.5 ± 22.4	467.8 ± 12.7	1456.4
p-Hydroxybenzoic acid	23.1 ± 2.5	60.3 ±5.4	81.2 ±9.4	568.0 ± 18.7	732.6
m-Hydroxybenzoic acid	10.8 ± 0.6	ND	157.0 ± 4.5	37.1 ± 4.2	204.9
Gentisic acid	1.8 ± 0.2	ND	ND	ND	1.8
Syringic acid	54.8 ± 1.2	433.5 ± 16.0	151.8 ± 8.5	710.3 ± 25.6	1350.4
Salicylic acid	60.7 ± 7.1	1.5 ± 0.5	ND	42.5 ± 5.7	104.7
p-Coumaric add	162.7 ± 12.9	1284.2 ± 17.5	1354.1 ± 15.2	596.6 ± 22.3	3397.6
Caffeic add	310.6 ± 21.3	1094.9 ± 12.4	1193.8 ± 17.3	1075.0 ± 5.4	3674.3
Ferulic acid	14.2 ± 1.8	69.9 ± 5.1	72.4 ± 7.0	53.7 ± 10.6	210.2
Sinapic acid	0.9 ± 0.2	0.9 ± 0.0	30.3 ± 0.9	41.0 ± 8.4	73.1
Trans-cinnamic acid	ND	ND	ND	6.6 ± 1.8	6.6
$\sum_{\text{benzoics}}$	210.1	949.3	871.5	1898.7	3929.6
$\sum_{\text{cinnamics}}$	488.4	2449.9	2650.6	1772.9	7361.8
$\sum_{\text{benzoics}}(\%)$	30.1	27.9	24.7	51.7	34.8
$\sum_{\text{cinnamics}}(\%)$	69.9	72.1	75.3	48.3	65.2
Total[d]	698.5	3399.2	3522.1	3671.6	11291.4

[a] Values are mean ± SD (n = 3).
[b] Total is the sum of each phenolic acid of four phenolic fractions.
[c] Not detected.
[d] Total is the sum of individual phenolic acids identified in each phenolic fraction.

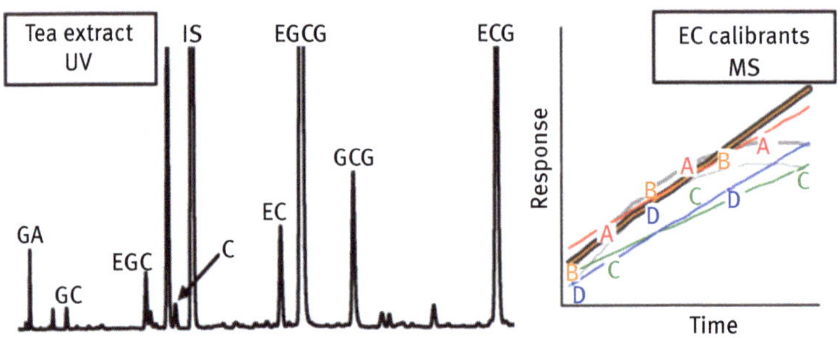

Figure 4.6: Schematic diagram of data acquisition by LC–UV and LC–ESI/MS.

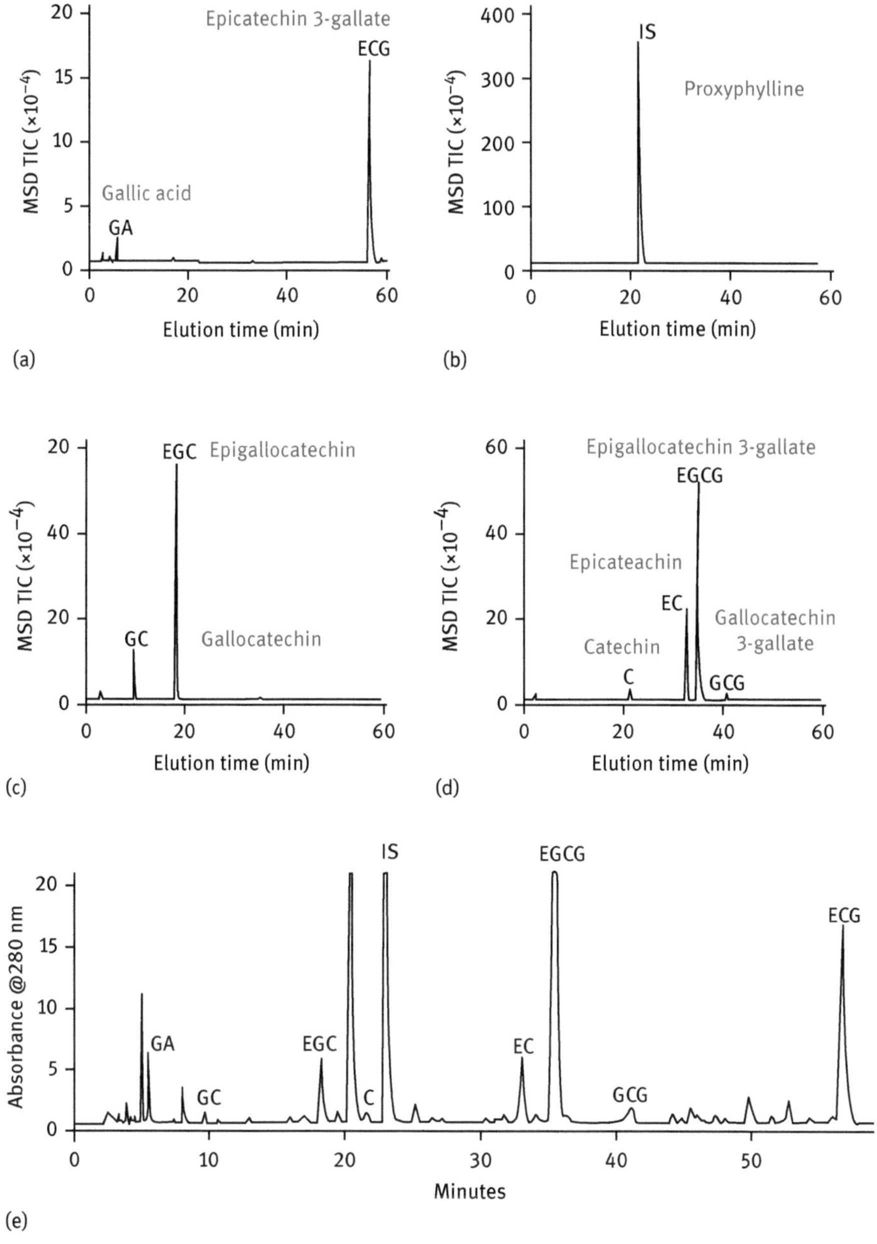

Figure 4.7: Antioxidants in tea separated by LC–MS spectra [16].

References

[1] Hollman P C H, Trijp J M P, Buysman M N C P. Fluorescence detection of flavonols in HPLC by postcolumn chelation with aluminum. Analytical Chemistry, 1996, 68(19): 3511–3515.
[2] Guyot S, Marnet N, Laraba D, et al. Reversed-phase HPLC following thiolysis for quantitative estimation and characterization of the four main classes of phenolic compounds in different tissue zones of a French cider apple variety (Malus domestica var. Kermerrien). Journal of Agricultural and Food Chemistry, 1998, 46(5): 1698–1705.
[3] Koleva II, Niederlander H A G, Beek T A V. An on-line HPLC method for detection of radical scavenging compounds in complex mixtures. Analytical Chemistry, 2000, 72(10): 2323–2328.
[4] Burns J, Gardner P T, O'Neil J, et al. Relationship among antioxidant activity, vasodilation capacity, and phenolic content of red wines. Journal of Agricultural and Food Chemistry, 2000, 48(2): 220–230.
[5] Fukumoto L R, Mazza G. Assessing antioxidant and prooxidant activities of phenolic compounds. Journal of Agricultural and Food Chemistry, 2000, 48(8): 3597–3604.
[6] Zheng W, Wang S Y. Antioxidant activity and phenolic compounds in selected herbs. Journal of Agricultural Food and Chemistry, 2001, 49(11): 5165–5170.
[7] Robbins R J. Phenolic acids in foods: An overview of analytical methodology. Journal of Agricultural Food and Chemistry, 2003, 51(10): 2866–2887.
[8] Tsao R, Yang R, Christopher J, et al. Polyphenolic profiles in eight apple cultivars using high-performance liquid chromatography (HPLC). Journal of Agricultural Food and Chemistry, 2003, 51(21): 6347–6353.
[9] Haghi G, Hatami A. Simultaneous quantification of flavonoids and phenolic acids in plant materials by a newly developed isocratic high-performance liquid chromatography approach. Journal of Agricultural Food and Chemistry, 2010, 58(20): 10812–10816.
[10] Chandra A, Rana J, Li Y Q. Separation, identification, quantification, and method validation of anthocyanins in botanical supplement raw materials by HPLC and HPLC-MS. Journal of Agricultural Food and Chemistry, 2001, 49(8): 3515–3521.
[11] Hernandez-Ledesma B, Davalos A, Bartolome B, et al. Preparation of antioxidant enzymatic hydrolysates from (alpha-lactalbumin and beta-lactoglobulin. Identification of active peptides by HPLC-MS/MS. Journal of Agricultural Food and Chemistry, 2005, 53(3): 588–593.
[12] Ayaz F A, Hayirlioglu-Ayaz S, Gruz J, et al. Strnad Separation, characterization, and quantitation of phenolic acids in a little-known blueberry (Vaccinium arctostaphylos L.) fruit by HPLC-MS. Journal of Agricultural Food and Chemistry, 2005, 53(3): 8116–8122.
[13] Nagy K, Courtet-Compondu M C, Holst B, et al. Comprehensive analysis of vitamin E constituents in human plasma by liquid chromatography-mass spectrometry. Analytical Chemistry, 2007, 79(18): 7087–7096.
[14] Buendia B, Gil M I, Tudela J, et al. HPLC-MS analysis of proanthocyanidin oligomers and other phenolics in 15 strawberry cultivars. Journal Agricultural Food and Chemistry, 2010, 58(7): 3916–3926.
[15] Boulekbache-Makhlouf L, Meudec E, Chibane M, et al. Analysis by high-performance liquid chromatography diode array detection mass spectrometry of phenolic compounds in fruit of Eucalyptus globulus cultivated in algeria. Journal of Agricultural Food and Chemistry, 2010, 58(24): 12615–12624.
[16] Bedner M, Duewer D L. Dynamic calibration approach for determining catechins and gallic acid in green tea using LC–ESI/MS. Analytical Chemistry, 2011, 83(16): 6169–6176.
[17] Hao M, Beta T. Qualitative and quantitative analysis of the major phenolic compounds as antioxidants in barley and flaxseed hulls using HPLC/MS/MS. Journal of the Science of Food and Agriculture, 2012, 92(10): 2062–2068.

[18] Aaby K, Ekeberg D, Skrede G. Characterization of phenolic compounds in strawberry (Fragaria x ananassa) fruits by different HPLC detectors and contribution of individual compounds to total antioxidant capacity. Journal of Agriculture and Food Chemistry, 2007, 55(10): 4395–4406.
[19] Bayram B, Esatbeyoglu T, Schulze N, et al. Comprehensive analysis of polyphenols in 55 extra virgin olive oils by HPLC-ECD and their correlation with antioxidant activities. Plant Foods for Human Nutrition, 2012, 67(4): 326–336.
[20] Castro-Gamboa I, Cardoso C L, Silva D H S, et al. HPLC-El CD: An useful tool for the pursuit of novel analytical strategies for the detection of antioxidant secondary metabolites. Journal of the Brazilian Chemical Society, 2003, 14(6): 771–776.
[21] Carrasco-Pancorbo A, Cerretani L, Bendini A, et al. Evaluation of the antioxidant capacity of individual phenolic compounds in virgin olive oil. Journal Agricultural Food and Chemistry, 2005, 53(23): 8918–8925.
[22] Zettersten C, Co M, Wende S, et al. Identification and characterization of polyphenolic antioxidants using on-line liquid chromatography, electrochemistry, and electrospray ionization tandem mass spectrometry. Analytical Chemistry, 2009, 81(21): 8968–8977.
[23] Zhang Haiying, Xue Jie. Hawthorn extract of total flavonoids. Xinjiang Medicine (Chinese Journal), 2006, 24(5): 17–19.
[24] Tao Yongyuan, Cai Chenbo, Shu Kangyun, et al. Decompression sublimation method to extract tea polyphenols from low-grade green tea. Henan Agricultural Sciences (Chinese Journal), 2012, 41(5): 53–55.
[25] Wang Yanfeng, Li Yanqing, Hao Yonghong, et al. Ultrasonic extraction of flavonoids from Ginkgo biloba. Food Science (Chinese Journal), 2002, 23(8): 166–167.
[26] Guo Zijie, Huang Ruqiang, Effect of Microwave on Leaching of Saponin in Panax Notoginseng, Chinese herbal medicine (Chinese Journal), 2007, 30(02): 232–234.
[27] Liu Yuan, Hou Binbin, Li Nan. On the extraction of flavonoids from Trollius. Food Research and Development (Chinese Journal), 2005, 26(06): 33–37.
[28] Jin Rucheng, Zhao Guolei, Jia Chau, et al. Supercritical CO2 extraction of red clover isoflavones technology research. Modern China Pharmacy (Chinese Journal), 2008, 25(7): 625–628.
[29] CAI Ding-guo, MIAO Ping, GU Ming-juan. High-speed countercurrent chromatography was used to isolate isorhamnetin, kaempferol and quercetin from Ginkgo biloba. Chinese medicine and clinical pharmacology (Chinese Journal), 1999, 10(01), 44–44.

5 Determination of antioxidant capacity by the electrochemical method

Electrochemical analysis is an important part of modern instrumental analysis. It is a qualitative and quantitative instrumental analysis of components. It is based on the electrochemical properties of the solution and its changing laws and is built on the basis of the relationships between electrical quantities such as potential, conductance, current, and electricity, as well as the measured relationships between certain quantities of substances. Because of its low instrumental cost, simple operation, high sensitivity, low detection limit, and simple sample preparation, it has gained considerable interest among antioxidant researchers. Based on whether free radicals are involved in the determination of antioxidant capacity, it can be divided into electrochemical methods involving nonfree radicals and electrochemical methods involving free radicals.

5.1 Electrochemical methods involving nonfree radicals

Nonfree radicals involved in the electrochemical method refer to the antioxidant capacity in the determination of the process. On the basis of the nature of antioxidants (such as redox) used for determination, the main methods are potentiometry and voltammetry.

5.1.1 Potentiometric determination of antioxidant capacity

The potential analysis method is based on the measured electrode potential based on the Nernst formula to determine the concentration of the substance to be tested. The Nernst formula is as follows:

$$\mathrm{ox} + ne \rightarrow \mathrm{red}$$

$$E = E^{\ominus} + \frac{RT}{nF}\mathrm{In}\frac{C_o}{C_R}$$

where E is the electrode potential, $E^{\ominus}$ is the standard electrode potential of the redox species, R is the molar gas constant, T is the thermodynamic temperature, n is the number of electrons transferred during the electrode reaction, F is Faraday's constant, C_O is the oxidizing species, and C_R is the concentration of the reducing substance.

Currently, the absolute potential of a single electrode cannot be measured. The measurement generally requires two electrodes to form a current loop (Figure 5.1),

https://doi.org/10.1515/9783110573763-005

where one of the electrodes is an indicator electrode whose electrode potential varies with the concentration to be measured and the other electrode is a reference electrode whose potential is not affected by changes in the concentration of the test solution. However, taking into account the effect of solution resistance during the measurement of the potential, the Luggin capillary tube is often used in the measurement process, making the reference electrode closer to the working electrode.

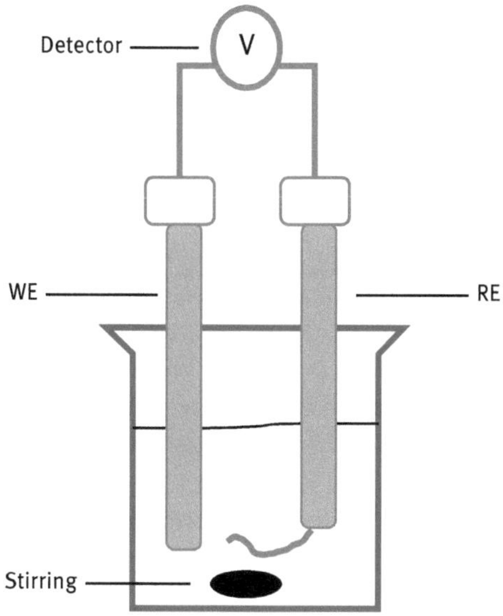

Figure 5.1: The device used for potentiometric method.

The potentiometric method is used for the detection of antioxidant capacity. The method has following advantages: easy operation, low cost, and strong anti-interference ability (especially for colored materials) [1–5]. Potentiometric determination of antioxidant capacity requires a media. Thus, we must first select a suitable media, which must meet the condition that the oxidation–reduction potential exists between the oxidation of active substances and antioxidants, which is as follows:

$$E^{\ominus}_{\mathrm{Rad/Rad(red)}} > E^{\ominus}_{\mathrm{ox/red}} > E^{\ominus}_{\mathrm{AOox/AO_{red}}}$$

where $E^{\ominus}_{\mathrm{Rad/Rad(red)}}$ is the oxidation–reduction potential of the active material; $E^{\ominus}_{\mathrm{ox/red}}$ is the redox potential of the mediator; and $E^{\ominus}_{\mathrm{AOox/AO_{red}}}$ is the redox potential of antioxidants.

Generally, the ROS (reactive oxygen species) redox potential is higher, that is, E = 1.55 V (vs SHE), while the common redox potential of antioxidants in the range of −0.42 to +0.30 V. Potassium ferricyanide/potassium ferrocyanide ($K_4[Fe(CN_6)]/K_3[Fe(CN_6)]$) is often chosen as the intermediate mediator due to its appropriate redox potential [6–10]. The basic principle of the potentiometric determination of antioxidant capacity is as follows:

$$E = E_0 + b\lg C_{ox}/C_{red}$$

$$E_1 = E_0 + b\lg(C_{ox} - X)/(C_{red} - X)$$

$$\text{AOA} = \pm(C_{ox} - \alpha C_{red})/(1 + \alpha)$$

where E is the redox potential of the antioxidant added to the presystem; E_0 is the standard redox potential of $K_4[Fe(CN_6)]/K_3[Fe(CN_6)]$; C_{ox} is the concentration of $K_3[Fe(CN_6)]$; C_{red} is the concentration of $K_4[Fe(CN_6)]$; E_1 is the redox potential of the antioxidant system added to the system; X is the concentration of antioxidant in the reaction; AOA is antioxidant capacity; $\alpha = 10^{(E1-E)/b}$; and $b = 2.3RT/nF$.

When antioxidants are added, the antioxidant reacts with the mediator, causing a change in the concentration ratio of the redox couple of the mediator. According to the Nernst equation, it can be seen that the change in the concentration can result in the change in potential, and the antioxidant capacity can be evaluated by measuring the potential change in the system. Brainina et al. used this method to determine the antioxidant capacity of antioxidants. There is a strong correlation between the test results obtained by the TEAC (Trolox Equivalent Antioxidant Capacity), electrochemical luminescence, and light colorimetric methods [10]. They also measured the actual samples, especially the red blood cell extracts gave good results, effectively eliminating the spectrum that often creates a problem of color interference. However, a challenge with potentiometry is that reactants or products get readily adsorb to the surface of the electrode, causing potential drift.

To reduce the pollution caused by the adsorption of the electrodes, the researchers designed a flow antioxidation capacity detection system. The flow system can quickly transfer the reactants or products on the electrode surface, thus keeping the electrode surface, as far as possible, clean. The system is simple to operate and can, to some extent, achieve high-throughput assays.

In addition to K_4 $[Fe(CN_6)]/K_3[Fe(CN_6)]$ that is selected as a vehicle, another commonly used mediator is a halogen pair such as Cl^2/Cl^-, I^2/I^-, and so on [3, 12]. The halogen pair has a faster reaction rate when it reacts with the oxidizing active substance and the antioxidant, thus reducing the time of the measurement. Let us consider $I^2/2I^-$ as an example. As shown in Figure 5.3, first, APPH is cleaved to generate free radicals AOO·. AOO· can oxidize I^- to elemental iodine. Antioxidants can affect the concentration of elemental iodine. The change in the potential due to the change in the concentration of $I^2/2I^-$ was measured so as to determine the antioxidant capacity of the antioxidant. The method is simple and requires few reagents, and can also be used for the analysis and detection of antioxidant capacity of high-flux antioxidants.

The potentiometric method used to determine the antioxidant capacity of the antioxidant has the following advantages: simple equipment, inexpensive reagents, easy to operate, and can also be used for determining capacity for large quantities.

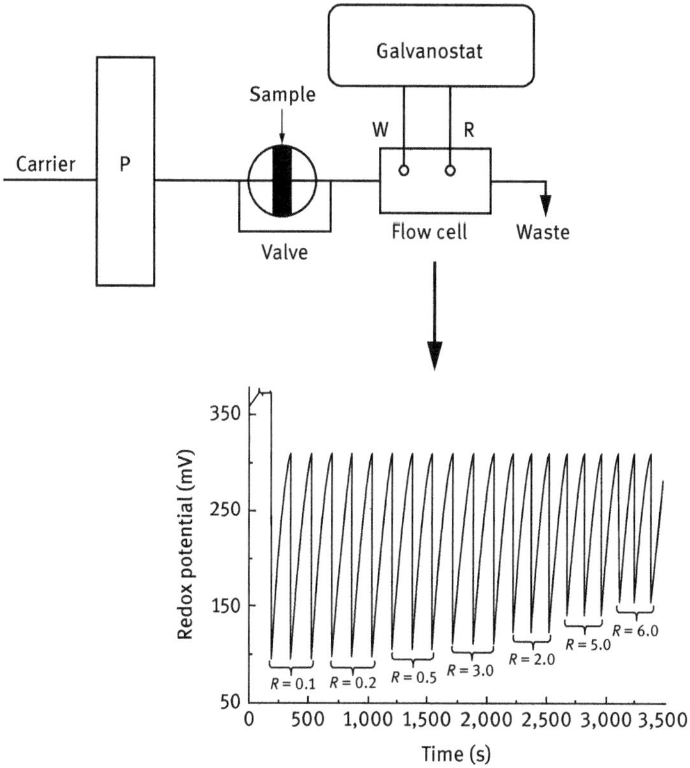

Figure 5.2: A schematic diagram of the flow system potentiometric device and flow measurement results.

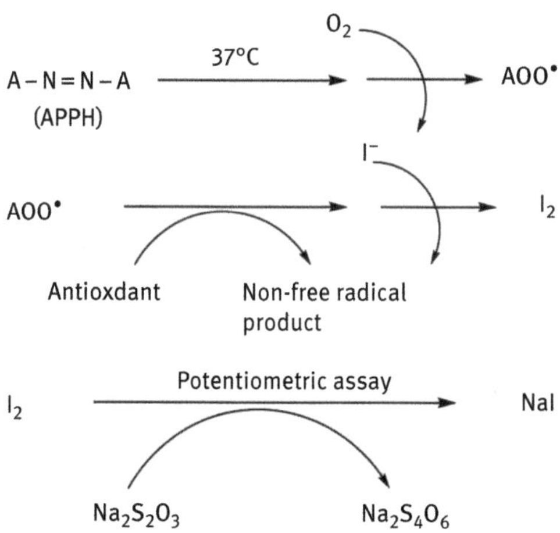

Figure 5.3: The titration potential principle of $I_2/2I^-$ [12].

However, the method has a few disadvantages: slow response, the electrode potential when working for a long time need to be corrected, and the working electrode can be easily adsorbed by the reaction material, resulting in poisoning.

5.1.2 Voltammetric determination of antioxidant capacity

Voltammetry is a special form of electrolysis. Based on the resulting current–voltage–curve analysis, it involves a small area of the working electrode and the auxiliary electrode to form the electrolytic cell. The most classic voltammetry is cyclic voltammetry (CV). CV results in a linear change on the working electrode voltage to record the current curve with the change in potential. The applied voltage is an equilateral triangle or equilateral staircase wave. A typical CV method is shown in Figure 5.4. When the applied electrode potential can cause a reduction of the species of oxide, the resulting cathode current is due to the following electrode process:

$$Red - e \rightarrow O_x$$

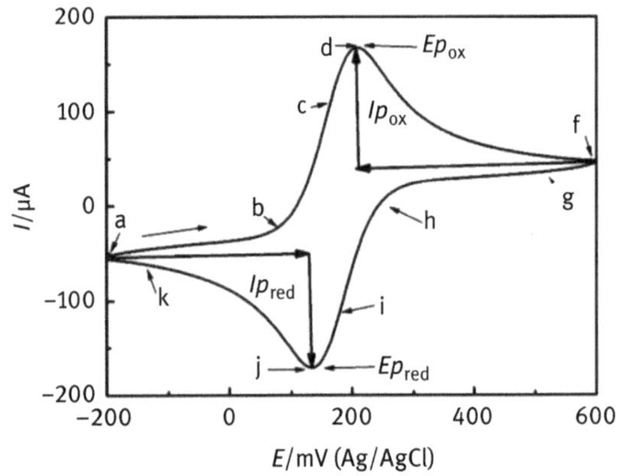

Figure 5.4: Cyclic voltammogram of 1 mmol/L $K_4[Fe(CN_6)]/K_3[Fe(CN_6)]$.

With increase in the applied potential, the oxidation current rapidly increases as well (*b–d*), decreasing the concentration of reduced species on the surface of the electrode close to zero. The current reaches its peak at the point *d* and then decays rapidly (*d–g*) as the "red" electrolyte in the solution near the electrode is converted to "O_x" and is depleted. For the reverse sweep, the potential sweep turns positive at *f*. However, the potential is still quite positive and can still oxidize the "red" species. Although the potential is swept in the negative direction, the red species accumulated near the electrode that diffuse to the electrode surface can still be oxidized by the electrode process. However, as the potential sweeps along the negative direction, the "O_x" species in the system gradually begins to revert to "red":

$$O_x + e \rightarrow Red$$

This generates a reduction current (i–k). The first cycle is completed when the potential reaches the starting potential E_i at the point a. Several important parameters can be obtained from cyclic voltammograms: anodic peak current (I_{pa}), cathodic peak current (I_{pc}), anodic peak potential (E_{pa}) and cathodic peak potential (E_{pc}). These parameters are important for the determination of the redox properties of the reactive species and their corresponding concentrations.

There are two main ways for antioxidants in the body to eliminate the free radicals and harmful substances: ① by preventing the generation of free radicals, which can chelate metal ions to inhibit the activity of free radical generating enzymes and ② by the redox reaction to eliminate the freedom. The stronger is the reducibility of the antioxidant, the stronger is its ability to eliminate free radicals. As can be seen from Figure 5.4, CV can be obtained and the redox properties of the material can be determined from the graph of the reduction of potassium ferricyanide, peak current I_{Pred} and reduction peak potential E_{Pred}, and the reduction of potassium ferrocyanide, oxidation peak current I_{Pox} and oxidation peak potential E_{Pox}. The redox potential of potassium ferricyanide and potassium ferrocyanide is

$$E_{red/ox} = \frac{1}{2}\left(E_{P_{red}} + E_{P_{ox}}\right)$$

$E_{red/ox}$ reflects the material's redox properties. The smaller the value of $E_{red/ox}$, the greater is the reduction ability. From Figure 5.4 we can obtain the concentration of electroactive substances. From the formula, we can see

$$i_p = 269 A n^{3/2} D_R^{1/2} C_R v^{1/2}$$

where i_p is the peak current; A is the area of the electrode; D is the diffusion coefficient; C_R is the concentration of the reducing agent; and v is the potential scan rate.

For the specific analyte, since the diffusion coefficient, area of the electrode, number of electrons delivered, and scan rate are all constants, due to the proportional relationship between peak current and the analyte concentration, the concentration of the active substance can be calculated from the peak current. Because CV is a redox process, the concentration of the active material as measured by the CV method also reflects the redox potential of the active material. Similarly, the antioxidant CV can also reflect its ability to reduce the concentration [13, 17]. Figure 5.5(a) shows a CV plot of the antioxidant, catechin [18], and we can easily obtain its redox potential.

However, not all antioxidants have reversible redox peaks, and most antioxidants do not show any appreciable reduction peaks. At this point, $E_{pa}{}^{1/2}$ of the antioxidant is used to indicate its antioxidant capacity. According to the principle of

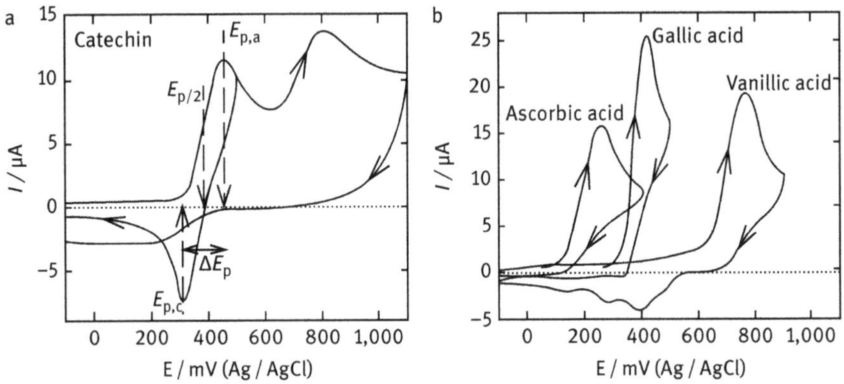

Figure 5.5: Cyclic voltammograms of theophylline (a) and ascorbic acid, gallic acid, and vanillic acid (b).

the aforementioned analysis, the reduction order of the three antioxidants can be obtained from Figure 5.5(b): ascorbic acid> gallic acid> vanillic acid. As the research progresses, the researchers used the integrated area Q below the oxidation curve instead of I_p. Q represents the total amount of charge in the redox process, and the more the charge that the antioxidant can provide, Q can more accurately reflect the antioxidant charge transfer [9]. In addition, the total charge Q is used as a result, allowing the CV method not only to measure the antioxidant capacity of individual antioxidants but also to measure the antioxidant capacity of the total antioxidants in the mixture and to reflect the synergistic effects between the antioxidants.

CV has following advantages: ① the CV method can measure the total antioxidant capacity of the mixture, and can reflect the synergistic effects between antioxidants; ② in this method the sample preparation is simple and rapid, without any need of expensive instruments and complex techniques; ③ this method can quickly distinguish a large number of samples; ④ the CV method meets the needs of physiological sensitivity test [20, 22]; ⑤ this can not only determine the water-soluble antioxidants but also can also measure fat-soluble antioxidants. Differential pulse voltammetry [20–22] (DPV) and square wave voltammetry (SWV) [23, 24] were also introduced for the determination of antioxidant capacity to further improve the sensitivity of the assay.

Irrespective of the electrochemical technology type, the choice of an appropriate electrode material for the determination of antioxidant capacity is the first factor that needs to be considered. It is known that the antioxidants demonstrate strong reactions on gold, platinum, and other noble metal electrodes. However, when the solvent contains methanol and ethanol, redox reactions occur on the noble metal electrode, which further make it difficult for the accurate determination of the antioxidant capacity. When compared with noble metal electrodes, the use of a glassy carbon electrode can reduce the impact of organic solvents, for example, methanol and ethanol. A carbon electrode is often used as a working electrode for the determination of antioxidants. Although the carbon electrode avoids the interference of the solvent, the problem of pollution caused by the adsorption of the antioxidant on the electrode surface cannot be solved. To obtain good reproducibility of the measurement results, the electrode surface needs to be updated. At present, the common electrode surface treatment method is the physical method, that is, the mechanical polishing of the electrode surface. There are researchers who use electrochemical methods to update the electrode surface. First, the electrode is raised to a higher potential to oxidize the material adsorbed. The potential is then reduced to a lower potential to restore the electrode, thereby restoring the active electrode surface. When an electrochemical method is used to measure the antioxidant capacity, sensitivity is yet another factor of particular concern. To increase the sensitivity of the measurement, a number of nanomaterials have been used to modify electrode surfaces such as multiwalled carbon nanotubes [25–26], graphene, activated carbon, and various other carbon materials as well as composite materials of precious metals [27, 28]. Nanomaterials have higher specific surface area, which can effectively increase the active area of the electrode surface and, to a certain extent, can improve the sensitivity of antioxidant detection.

The electrode is adsorbed by the reactants and byproducts, resulting in poisoning the electrode surface, which is the biggest challenge while using electrochemical methods. At present, the best solution to the problem of electrode contamination is to use a flow system [30–35]. A flow system can reduce the electrode surface pollution for three main reasons: ① products can quickly flow out to reduce its concentration on the electrode surface; ② using microfluidic technology, less injection volume can be result in less oxidation products, thereby reducing pollution; and ③ reducing the potential can reduce the pollution of antioxidant products. Many researchers use a flow system to detect the antioxidant capacity of antioxidants, while avoiding the contamination of the electrodes to some extent. The flow system is shown in Figure 5.6, with an additional procedure to better maintain the electrode surface clean. After the measurement, the electrode is automatically cleaned for 1–2 min. The flow system was also designed as a 96-well cell for determining a large number of samples [29].

Raquel et al. measured the antioxidant capacity of antioxidants in combination with microfluidic technology (Figure 5.7) [36]. The method has the following advantages: ① the electrode potential is controlled at the common oxygen-active material potential so that the determination result can best simulate the actual antioxidant capacity of the antioxidant; ② the injection volume is small oxidation products and thus reduce pollution; ③ simultaneous determination of the synergistic effect of mixed antioxidants; and ④ the device is easy to integrate and can achieve high-throughput detection.

Currently, there are a number of methods for in vitro determination of the antioxidant capacity of antioxidants, and the measured in vitro and in vivo results have some differences. Therefore, the determination of the antioxidant capacity of antioxidants in

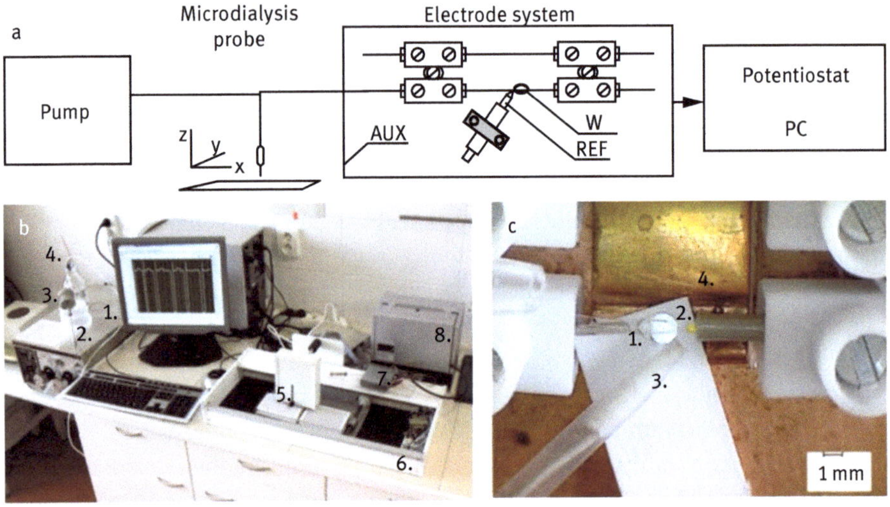

Figure 5.6: Determination of antioxidants in wine antioxidant capacity array electrode schematic (a); flow system device (b); and flow system electrode outlet (c).

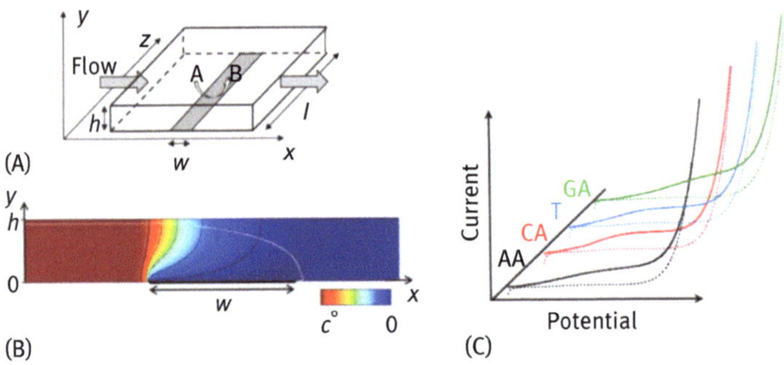

Figure 5.7: A schematic diagram of (a) three-dimensional flow microstrip electrode (a); two-dimensional concentration of liquid flowing through the microstrip electrode (b); and a linear scan of the four antioxidants (c).

vivo is of great significance in guiding nutrition intake and medical research. In electrochemical assays, electrodes that are prepared using special methods are microns or even nanometers in size, making it possible to electrochemically determine the antioxidant capacity of tissues and cells [37–40]. As shown in Figure 5.8, the researchers used microelectrodes to measure intracellular oxygen-active substances. The addition of antioxidants affected the content of the intracellular oxidized substances. Therefore, the microelectrodes could be used to measure the oxygen-active substances and then the antioxidant capacity of antioxidants. Determination of antioxidant capacity in vivo can result in large amount of in situ data, and then a database can be

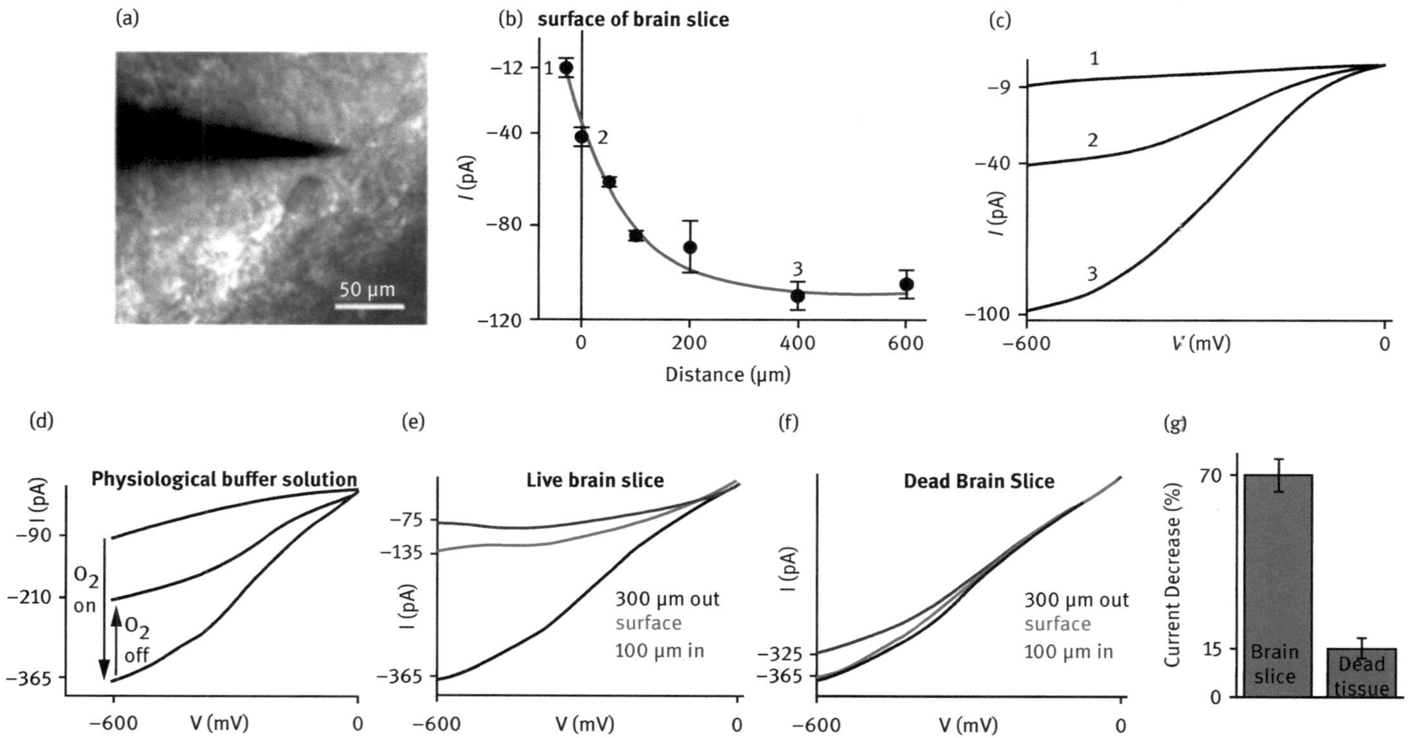

Figure 5.8: Determination of oxygen content inside and outside of brain slices: (A) the microelectrode in the brain section; (b) the current of the microelectrode and its position in the brain section at −600 mV (vs Ag/AgCl); (c) 1 – internal, 2 – surface, 3 – voltammogram of microelectrodes at different positions from −400 µm; (d) platinum-coated microelectrodes voltammogram under different oxygen conditions in physiological environment; (e) voltammogram of platinum-plated microelectrodes in living brain sections; (f) voltammogram of platinum-coated microelectrodes in dead brain sections: at 600 mV (vs Ag/AgCl), in live; and dead brain sections, the current drops [37].

built. Establishing a relationship between in vivo and in vitro data to replace in vivo assays with in vitro assays is of great importance for the determination of the antioxidant capacity of antioxidants.

5.1.3 Determination of antioxidant capacity by electrochemiluminescence

Electrochemiluminescence is a combination of chemiluminescence and electrochemistry. It refers to the luminous phenomenon caused by the electrochemical reactions, which is triggered by applying a certain voltage on the electrode surface. During the process, some biomass generated via electrochemical reactions, and then the excited state forms via electron transfer between some of these biomass or these biomass and system components. When the excited state returns to the ground state, luminous phenomenon occurs.

Electrochemiluminescence immunoassay has been widely used in biology, medicine, pharmacy, clinical, environmental, food, immunology, nucleic acid hybridization analysis, and industrial analysis because of its high sensitivity, simple equipment required, and easy operation and automation. In electrochemiluminescence research, the most crucial step is to find a suitable oxide material. The oxide material can be the electron transfer between the electrodes, generating luminescent substances or intermediates. Most used oxidative substances are polycyclic aromatic hydrocarbons, nitro compounds, luminol and bipyridyl ruthenium complexes, and sometimes we also need luminous bodies such as tripropylamine (TPA). Among these materials, the luminol-hydrogen peroxide and bipyridyl ruthenium-tripropylamine systems are often used. Tu Yifeng research group at Suzhou University successfully determined the antioxidant capacity by using the luminol-hydrogen peroxide system [41]. In recent years, Xiao Dan's, Sichuan University, research group has also proposed a new electrochemical light-emitting system – carbonitride (GCN)-TPA system. By using this system they could successfully determine antioxidant capacity of rutin [42].

As shown in Figure 5.9, GCN and TPA were oxidized to $C_3N_4^{+\cdot}$ and $TEA^{+\cdot}$, respectively, at a given applied potential, whereas $TEA^{+\cdot}$ lost hydrogen ions and further generated TEA-free radicals. $TEA^{\cdot}$ and $C_3N_4^{+\cdot}$ combine to form excited state $C_3N_4^{*}$, which will further return to ground state to generate stable GCN with emitting 470 nm light.

$$C_3N_4 - e \rightarrow C_3N_4^{\cdot+}$$

$$TEA - e \rightarrow TEA^{\cdot+}$$

$$TEA^{\cdot+} - H^+ \rightarrow (EtO)_2N^{\cdot}CHCH_2OH\ (TEA^{\cdot})$$

$$C_3N_4^{\cdot+} + TEA^{\cdot} \rightarrow C_3N_4^{*} + TEA\ oxidant$$

$$C_3N_4^{*} \rightarrow C_3N_4 + hv$$

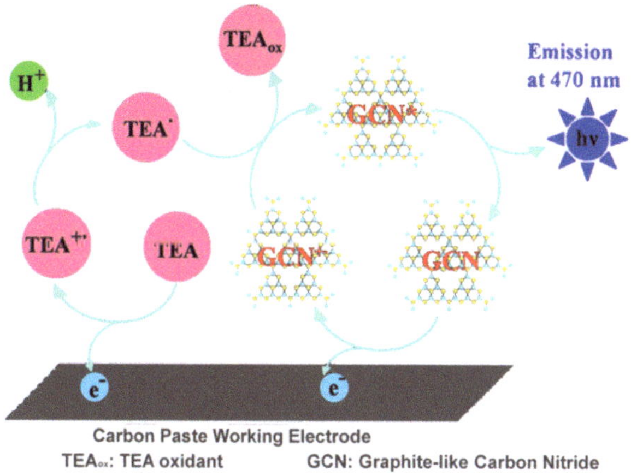

Figure 5.9: Electrochemical luminescence based on GCN-TPA system.

When rutin is added to the aforementioned system, rutin oxide and GCN can result in effective energy resonance transfer quenching and luminescence. On the basis of this principle, one can determine the antioxidant capacity of rutin.

The biggest advantage of electrochemical luminescence is the high sensitivity. In addition, the method is simple and fast, the result is stable, the error is small, and the device is easy for miniaturization.

5.1.4 Enzyme involvement method to determine antioxidant capacity

Phenol is one of the main antioxidants in food. Researchers often use the content of total phenols to measure the content of antioxidants. Therefore, the determination of total phenols has attracted a great deal of interest among researchers. However, there is a class of organisms called enzymes that specifically respond to phenols. Combined with electrochemical advantages and enzyme selectivity, researchers can use the enzyme biosensor methods to determine phenols [43–47].

Currently, the most used enzymes are tyrosinase and laccase [48–53]. These are two enzymes with copper ions as the active center. The divalent copper ions can oxidize phenols to corresponding ketones. Then Cu^+ will lose an electron to generate new Cu^{2+} ions and the next cycle begins. Some kinds of protein with certain spatial structures show selective distinctions toward a category of phenols, which finally realizes the determinations of these phenols in the system Figure 5.10.

Other enzymes such as HRP [54–55], peroxidase [86–87], polyphenol oxidase [58,59], glucose dehydrogenase [60, 61], cellobiose dehydrogenase [62], and so on can also be used for the detection of phenols. To improve the sensitivity of the detection, researchers also used a double-enzyme amplification system [63, 64]. As shown

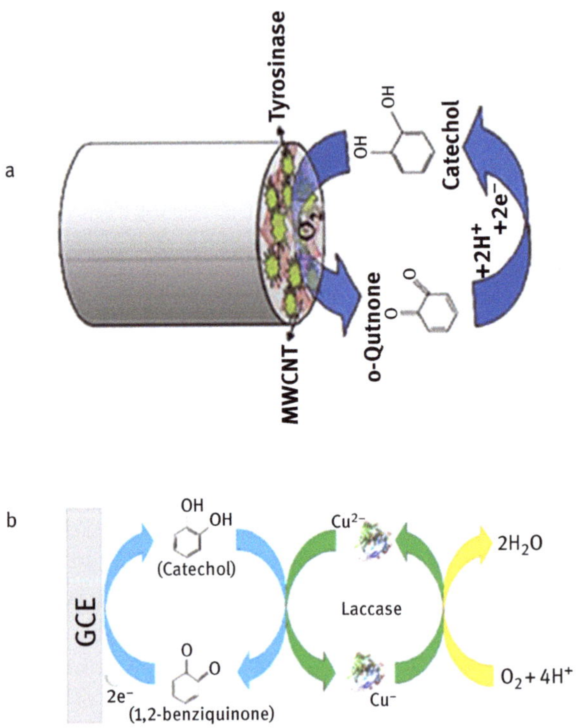

Figure 5.10: Tyrosinase (a) and laccase (b) modified electrode for the determination of o-diphenol.

in Figure 5.11, hydroquinone can be oxidized by laccase to p-benzoquinone while oxygen is reduced to water, and the resulting p-benzoquinone can be regenerated to hydroquinone and hydrogen during catalysis of QH-CDH enzyme ion. The diphenol content can then be determined by measuring the content of oxygen or hydrogen ions. When compared with a single-enzyme system, a double-enzyme system was used so it could improve the sensitivity of hydroquinone by 3–4 orders of magnitude. The lowest limit of detection of *p*-aminophenol and epinephrine was 70 pmol/L and 1 nmol/L, respectively.

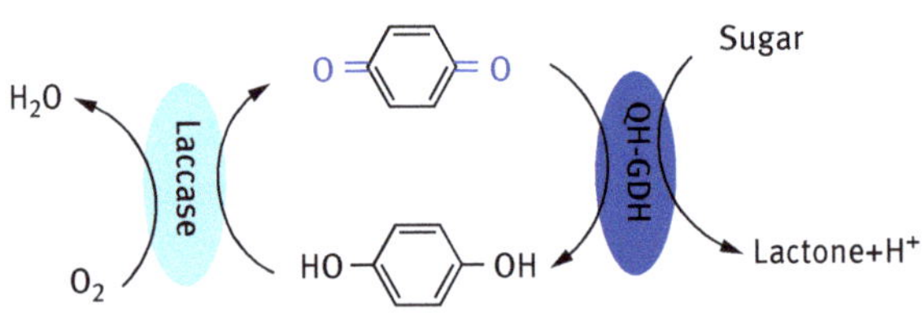

Figure 5.11: A double-enzyme system for the determination of hydroquinone.

As with other electrochemical methods, the surface of the modified electrode can still be easily poisoned. To solve the problem of the surface of the modified electrode being easily contaminated, Merkoci et al. [65] immobilized tyrosinase on the surface of the electrode and designed a microfluidic chip that could on-line determinate phenols in situ. The device is sensitive, fast, and reproducible (Figure 5.12). The biggest advantage of using the enzyme to determine the phenolic antioxidant capacity is that it can selectively measure the kind of phenol and lay a good foundation for the determination of the antioxidant fingerprint with the electrochemical method. However, the enzyme is expensive, difficult to preserve, and has severe conditions of use, thus limiting its widespread use.

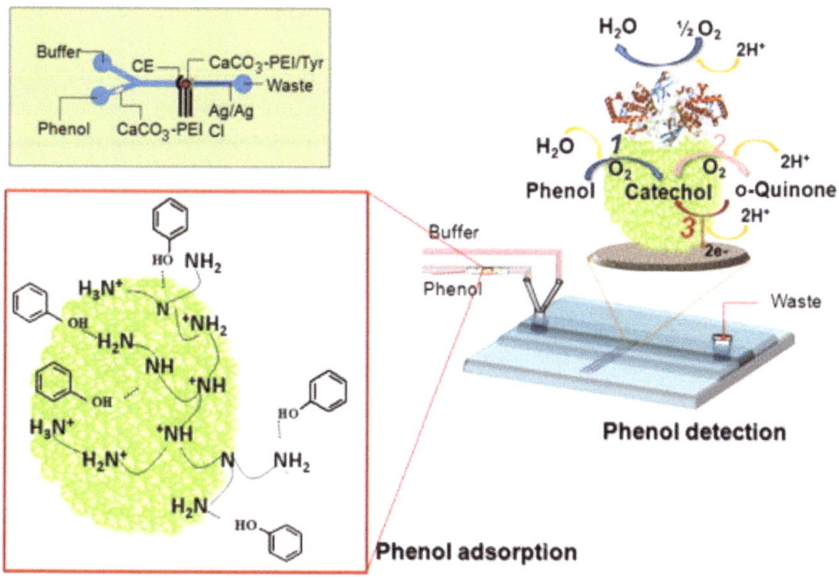

Figure 5.12: Microfluidic tyrosinase-modified electrode and the schematic diagram of the determination of phenols. The right image shows an enlarged view of the phenolic adsorption.

5.2 Electrochemical methods of free radical participation

5.2.1 Method for generating free radicals

The most common free radicals of oxidative active substances in vivo are ·OH and $O_2^{\cdot-}$. These free radicals are the main factors that damage the living macromolecules in vivo. Therefore, the determination of antioxidant capacity using ·OH and $O_2^{\cdot-}$ these two free radicals will be almost similar to the actual situation in vivo. At present, the electrochemical methods commonly used for the determination of antioxidant capacity generation of free radicals are as follows:

1 ·OH production method

(1) Fenton reaction

The Fenton reaction is a classic chemical free radical generation method. It generates ·OH by the reaction of reduced transition metal ions such [66, 67] as Fe (II), Cu(I), Cr (II), and H_2O_2, The specific process is as follows:

$$M_e^n + H_2\ O_2 \rightarrow Me^{n+1} + OH^- + \cdot OH$$

where M^n and M^{n+1} are transition metal ions. The main factors affecting the production of OH are pH, reaction temperature, H_2O_2, and transition metal ion concentration. Therefore, to produce ·OH of ideal concentration, we must satisfy the aforementioned conditions reasonably.

(2) Photochemical methods

Photochemical production of ·OH is based on the nature of the semiconductor. A semiconductor is divided into the conduction band and valence band. In the light of the appropriate wavelength of radiation, resulting in e^- and h^+, the position of the conduction band determines the e^- reduction, and the position of the valence band decides the oxidation of h^+, where appropriate oxidation of h^+ is used to oxidize water to produce ·OH [68, 69]. The following is a description of the semiconductor TiO_2 (Figure 5.13).

The chemical reaction is as follows:

$$TiO_2 + hv \rightarrow h_{VB}^+ + e_{CB}^-$$

$$h_{VB}^+ + H_2O_{ads} \rightarrow \cdot OH + H^+$$

$$h_{VB}^+ + 2OH_{ads}^- \rightarrow OH^- + \cdot OH$$

$$e_{CB}^- + O_2 \rightarrow O_2^{-\cdot}$$

$$O_2^{-\cdot} + 2H_2O \rightarrow 2 \cdot OH + 2OH^- + O_2$$

where hv, h^+_{VB}, and e^-_{CB} are ultraviolet light, holes, and electrons, respectively.

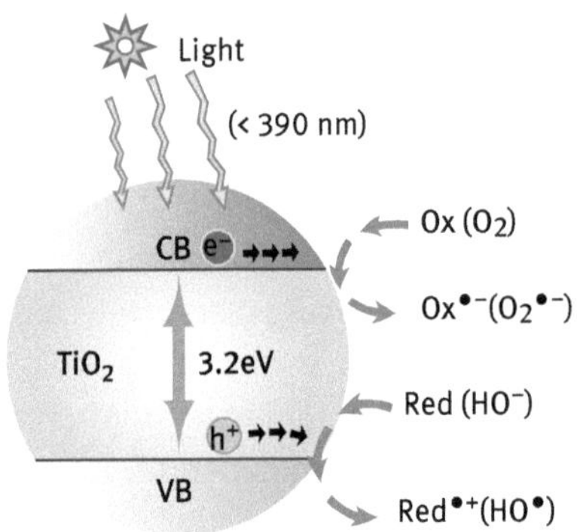

Figure 5.13: Oxygen-active species with TiO_2 in light irradiation.

In the process of the irradiation of ultraviolet light, TiO_2 is excited to produce e^- and h^+, where h^+ can oxidize water and hydroxyl groups on the adsorption surface to generate hydroxyl radicals; on the other hand, valence band electrons can react with oxygen to generate superoxide free radicals. These short-lived superoxide radicals can again react with water to generate a large number of hydroxyl radicals.

When compared with the conventional chemical methods, optical methods have the following advantages: easy control, convenient operation, and reusable catalysts, which are especially beneficial for the integration of instrumentation systems. Therefore, more researchers focus on the optical methods.

(3) Electrochemical methods to generate free radicals
Precious metals and their nanoparticles produce hydroxyl radicals on catalytic reduction of oxygen. Liu et al. [70] modified palladium oxide nanoparticles on the surface of the electrode and scanned the potential in a range of 0.8–0.8 V for 1500 cycles in a solution containing hydrogen peroxide or oxygen, eventually detecting hydroxyl radicals in the solution. Therefore, the electrochemical method can be used for generating hydroxyl radicals. At the same time, the generation of hydroxyl radicals can be effectively controlled based on the regulated potentials.

2 O_2^- production method

The classic O_2^- production method is the chemical method. In xanthine oxidase (XOD) catalysis, xanthine is oxidized to uric acid, while generating O_2^- [71]:

$$Xahine + O_2 + H_2O \overset{XOD}{\rightarrow} Uric\ acid + 2H^+ + O_2^{-\cdot}$$

Similar to the Fenton reaction method, this method also needs to adjust the concentration of various reactants to obtain the optimal concentration of O_2^-, and the conditions are more severe due to the participation of the biological enzyme. The addition of NaOH to DMSO can also generate O_2^-, or O_2^- can also be produced by adding KO_2 to DMSO [72]. The biggest challenge for generation of O_2^- via chemical methods lies in the difficult experiment conditions to be controlled.

5.2.2 Antioxidant capacity based on DNA damage

The biomolecules that hydroxyl radicals attack primarily in the body are DNA, which oxidizes DNA and further breaks the DNA bimolecular phosphate backbone [73, 74]. Among the four bases A, T, C, and G that consist DNA, the G base has the lowest redox potential and is the most vulnerable to attack by hydroxyl radicals. Therefore, the G base is first selected for the antioxidant capacity detection [75, 76]. Figure 5.14(a) [77] shows the process of measuring the antioxidant capacity using a method of electrochemical DNA damage. First, the G base is immobilized on the surface of the glassy carbon electrode. SWV is used to measure the oxidation signal of the G base because G higher base concentration leads to higher oxidation current. ① When the G base is oxidized in the presence of hydroxyl radicals, the concentration of the G base decreases. As a result, the oxidation current is reduced. ② When antioxidants (such as ascorbic acid) are introduced, the antioxidant has a low redox potential and can react with the hydroxyl radical in the G base to protect the G base. Then, the oxidation current of the G base increases. ③ Based on this principle we can measure the antioxidant

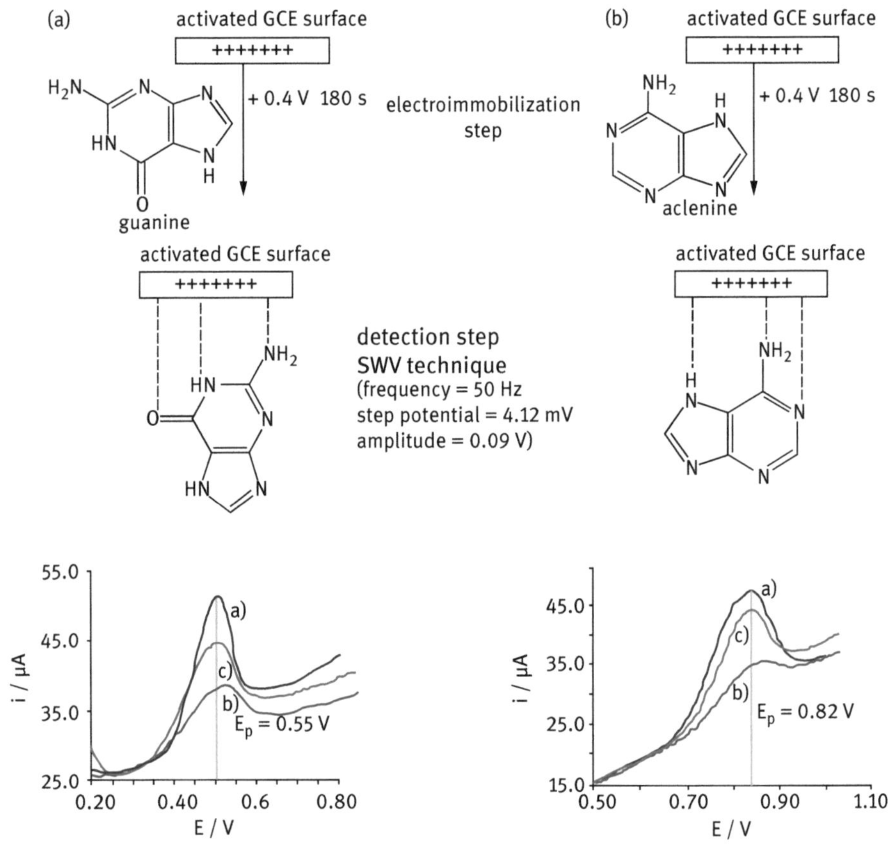

Figure 5.14: Electrochemical methods immobilize SWN maps of purine bases (G bases (a) and A bases (b)) on glassy carbon electrodes and in PBS (pH = 4.8): (a) a blank signal; (b) an electrochemical signal after hydroxyl radical damage; and (c) an electrochemical signal that is damaged in hydroxyl radicals by the addition of ascorbic acid.

capacity of antioxidants. In addition to G bases, other bases can also be used for the detection of antioxidant capacity (Figure 5.14(b)) [78, 79].

The disadvantage of using this method is that the antioxidant itself has a low redox potential and can easily interfere with the assay by reacting on the electrode. Another disadvantage is that a single base, after all, is not a DNA molecule and does not optimally mimic the true antioxidant capacity of the antioxidant.

To determine the results for the human body, the researchers used natural double-stranded DNA instead of DNA bases. The determination principle was similar to that of the G base. Commonly used natural DNA was calf thymus and herring sperm DNA. To increase [78, 80–83] the sensitivity of DNA-modified electrode signals, Liu et al. introduced Ru $(bpy)_3^{2+}$ mediator into the assay and successfully determined the antioxidant capacity of several antioxidants using kinetic methods. Regardless of the direct determination of DNA oxidation signal or use Ru $(bpy)_3^{2+}$ medium-assisted strategy, the applied potential was high. This may lead to the probable direct reaction of antioxidant on the electrode surface, which will affect the accurate measurement of antioxidant capacity to some extent. To solve this problem, methylene blue (MB) molecule is used to label double-stranded DNA. MB molecule has an oxygen reduction potential of −0.22 V (vs Ag/AgCl), which can effectively reduce the

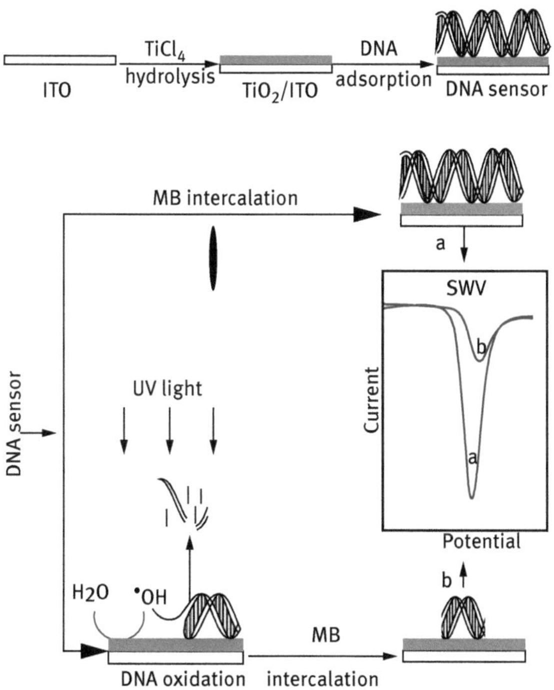

Figure 5.15: Optical methods used for producing hydroxyl radicals for DNA damage and electrochemical detection.

interference of antioxidants themselves. As shown in Figure 5.15 [84], for the modification of TiO_2 electrode-fixed double-stranded DNA, a large number of MB molecules can be combined with double-stranded DNA. By using SWV we can get MB of strong reduction current. In some cases, TiO_2 is triggered to generate a large amount of hydroxyl radicals. DNA by hydroxyl free radical will reduce the amount of MB molecules, reducing electrochemical signals. In the cases under illumination, TiO_2 is triggered to generate a large amount of hydroxyl radicals. DNA is damaged by hydroxyl free radical and cause the decreasing of MB combination molecules, which expresses in the decreasing of electrochemical signals. By introducing antioxidants into the system, they can protect DNA from damage and maintain the electrochemical current of MB. Based on this principle, the antioxidant capacity can be measured.

The advantages of using the DNA-modified electrode method to determine the antioxidant capacity of antioxidant are as follows: (i) by using DNA from the body's existing biological molecules, and choosing free radicals for the hydroxyl radical from the body's active oxygen free radicals, the results can be obtained to a good extent, simulating the real situation; (ii) this method can be used if the introduction of ultramicroelectrodes can be achieved in vivo to determine the antioxidant capacity of antioxidants, which is a topic for the future research.

5.2.3 Determination of antioxidant capacity based on ferritin including superoxide dismutase and cytochrome c

Superoxide free radicals are an important part of oxygen-active substances. Using superoxide radicals to measure antioxidant capacity of antioxidants will be a case very close to the true situation in a human body like DNA assay. Two proteases, superoxide dismutase (SOD) [85–87] and cytochrome c (Cty c) [88, 89], are the commonly used enzymes for the determination of superoxide radicals. SOD can be converted to O_2 and H_2O_2 based on the changes in the central metal atoms. SOD is an enzyme present in the body. Common SOD can be divided into four categories: Cu/Zn-SOD, Mn-SOD, Fe-SOD, and Ni-SOD. As in the case of other biological enzymes, the most crucial step for using SOD enzymes is to select a suitable enzyme immobilization method. At present, the commonly used enzyme immobilization methods for biosensors include adsorption, covalent bonding, cross-linking, embedding, gel method, and self-assembly method. To best maintain the biological activities of the biological enzyme, the most successful method among many enzyme immobilization methods is the self-assembly method. First select an appropriate electrode (gold electrode), and then the different carbon chain thiol molecules are fixed on the surface of the gold electrode based on the principle of complementary charge of SOD enzyme immobilization. In this method, the appropriate polymer molecules should also be selected for multilayer SOD enzyme loading on the surface of the electrode. To reduce the cost, a carbon fiber electrode can be used instead of the gold electrode to obtain the SOD enzyme load. As shown in Figure 5.16, first on the surface of the carbon fiber electrode gold nanoparticles get deposited and then cysteine is assembled. Based on the charged complementarity of cysteine and SOD, the SOD enzyme can be immobilized on the surface of the electrode, so that the SOD can retain the enzyme activity to the maximum, which can be measured by superoxide radical [90].

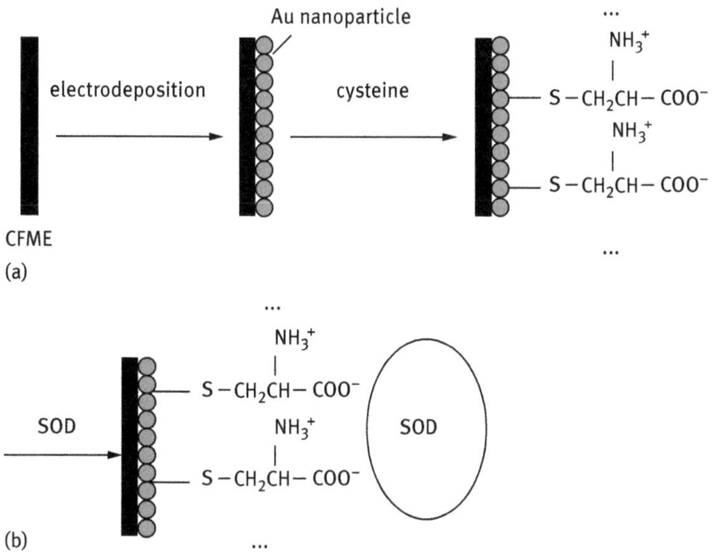

Figure 5.16: Electrodeposition of gold nanoparticles on the carbon fiber and the assembled cysteine process (a) and the process of immobilizing SOD enzyme (b).

Cty c can also determine superoxide radicals; Cty c is an iron-centered protein (Fe^{2+}/Fe^{3+}). Fe^{3+}-centered Cty c loses electrons in the electrode to generate Fe^{2+}-centered Cty c, and the latter can be oxidized by superoxide to generate Fe^{3+}-centered Cty c. On the basis of the aforementioned principle, we can measure superoxide radicals. As shown in Figure 5.17, Cyt c is immobilized on the surface of the electrode using the similar method to that of SOD. Cyt c can reduce the superoxide to oxygen to detect superoxide. The detection of superoxide is the first step in the analysis of antioxidant capacity [91]. By adding these two systems into the antioxidant, the antioxidant can compete with the superoxide, thus reducing the corresponding detection signal. Based on the change of the signal after adding the antioxidant, the antioxidant capacity of the antioxidant can be detected.

Using the method of enzymatic modification, the researchers attempted to measure the concentration of oxidized active substances in vivo. Tian et al. prepared ZnO microelectrodes (Figure 5.18). After the modification of SOD, the oxidized active substances in the soya bean sprouts were determined [92]. This method is simple. The electrode is easy to prepare and can be used for the detection and analysis of the antioxidant capacity of other plants' live superoxide and antioxidants.

In short, the electrochemical method for determining the antioxidant capacity of antioxidants is inexpensive and sensitive, and the sample does not require complicated pretreatment and can detect and analyze the antioxidant capacity of a single antioxidant and multiple antioxidants. However, when the electrochemical method is applied, the electrode is easy to be poisoned by the products and adsorbents. Also the modification process of biomolecule is complicated, while the reproducibility is

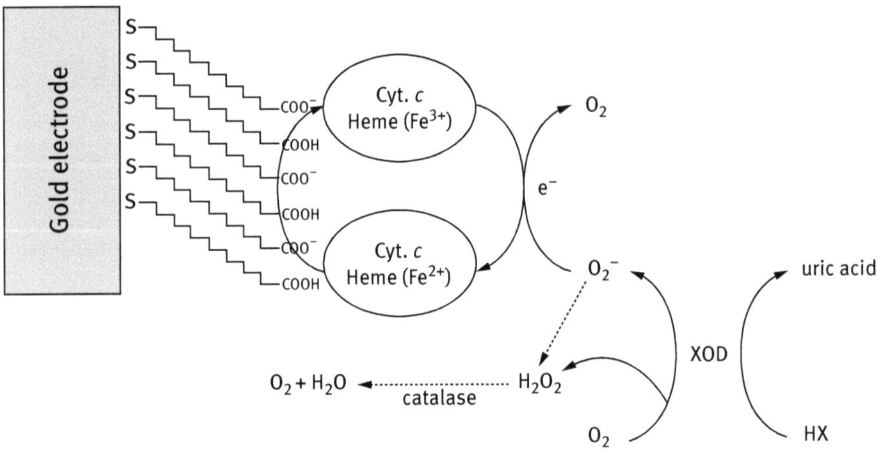

Figure 5.17: A schematic representation of the mechanism of superoxide generation in HX/XOD systems based on Cty c-modified electrodes.

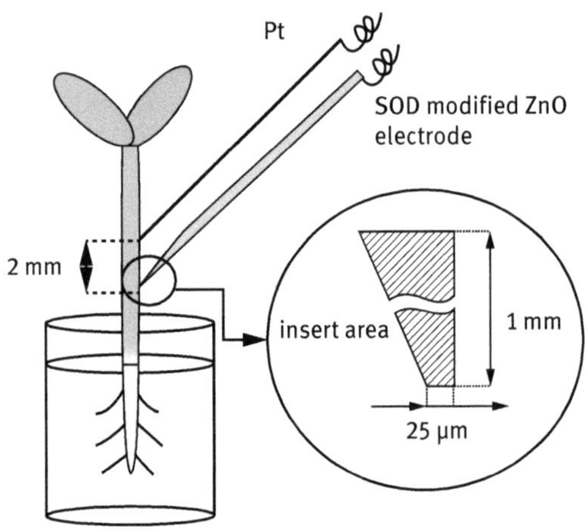

Figure 5.18: SOD-modified ZnO electrode for in situ determination of oxidative materials in plants [92].

poor and the measurement conditions are rigorous. All the above-mentioned problems need to be solved in the future.

5.3 Several examples of electrochemical measurement methods

The following is based on the current progress made in the field of electrochemical oxidation in China and across world to introduce three classic methods: the potential method of determination of antioxidant capacity, voltammetric determination of antioxidant capacity [18], and DNA damage determination [94] of antioxidant capacity [93].

5.3.1 Potentiometric determination of antioxidant capacity of antioxidant

Antioxidants in food play an important role in maintaining the health. The antioxidants in food (mainly for phenolic substances) can be eliminated by the oxidation of high-concentration active substances, thus helping in minimizing cardiovascular diseases, Parkinson's disease, and the risk of cancer. Therefore, assessing the total antioxidant capacity of antioxidants in food is of significant importance. There are many methods for the evaluation of antioxidant capacity; however, different methods show different assay results for the same antioxidant or mixed sample. In addition, some methods can only be used to measure water-soluble antioxidants, and some methods can be used only to measure fat-soluble antioxidants. The determination method varies depending on the measurement conditions, and different solvents or different pH values used in the same method show different measurement results. Therefore, for determining the antioxidant capacity of antioxidants, there is a need for a method that gives uniform and rapid results.

Owing to its high sensitivity, rapid response, and low cost, the electrochemical method has aroused widespread interest among researchers. However, the electrode gets easily poisoned, which is the greatest challenge. The main factor of poisoning is electrochemical reactants or products on the electrode surface adsorption. In the potential method, it is not required to apply potential on the surface of the electrode, and the surface adsorbate can be easily washed away by the flowing solution, which effectively solves the problem of electrode poisoning in the electrochemical measurement.

For potentiometric determination of antioxidant capacity, the researchers selected potassium ferricyanide and potassium ferrocyanide as probe molecules (Section 5.1.1). The measurement process is shown in Figure 5.19.

Under the action of a peristaltic pump, the sample solution (CS) and the probe molecule solution (RS) reach the reactor (RC) at a certain flow rate and reach the potential detector (FC) after the reaction. For the same sample, time taken for the sample solution and probe molecules to react in the reactor determines the signal

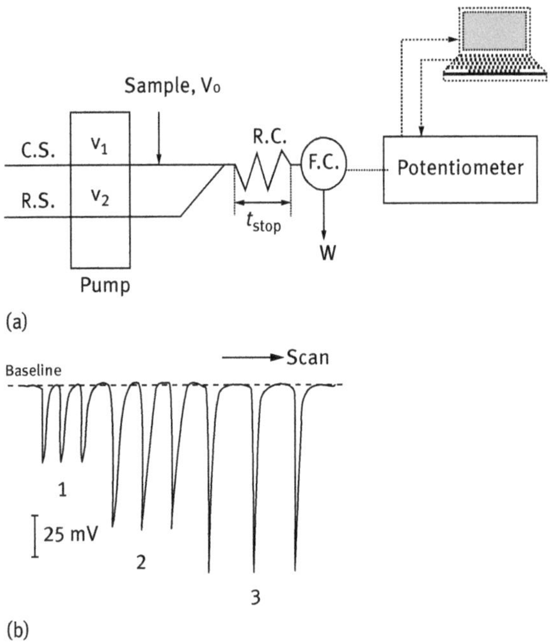

Figure 5.19: (A) Flow-injection potential method for the determination of antioxidant capacity. Antioxidant sample solution (CS), probe molecule $[Fe(CN)_6]^{3-}/[Fe(CN)_6]^{4-}$ (RS), reactor (RC), and potential detector (FC); (B) 0.5 mmol/L Asc and probe molecules 1.0 mmol/L $[Fe(CN)_6]^{3-}$/1.0 mmol/L $[Fe(CN)_6]^{4-}$ (pH = 7) potential response signal.

strength of the detector. Figure 5.19(b) shows the measured signal of antithrombotic (Asc) at different reaction times; it is obvious that the signal increases with the increase in reaction time.

The flow rate of the sample solution and probe molecule solution and the length of the reactor are important factors that influence the signal to be measured and need to be first optimized. Asc as the standard conditions for the optimization, 1 mmol/L of Asc injection volume of 30–500 μL, as shown in Figure 5.20(a); with the increase in the injection volume, the potential signal increases (400 μL). At 400 μL, the length of the reactor was simultaneously optimized and the signal increased with increasing length from 30 to 360 cm. Figure 5.20(b) shows the optimization of reactant and probe molecular flow rates, where both solutions decrease with an increase in the flow rate, with a reactant flow rate of 0.6 mL/min and a probe molecule flow rate of 0.4 mL/min optimal. The initial ratio of potassium ferrocyanide to potassium ferrocyanide in the probe molecule also has a strong effect on the assay signal (Figure 5.21(a)), with the best signal at a ratio of 1.0/0.1. Therefore, the optimal detection conditions were as follows: the injection volume was 400 μL; the flow rates of the reactants and probe molecules were 0.6 and 0.4 mL / min, respectively; and the reactor length was 240 cm.

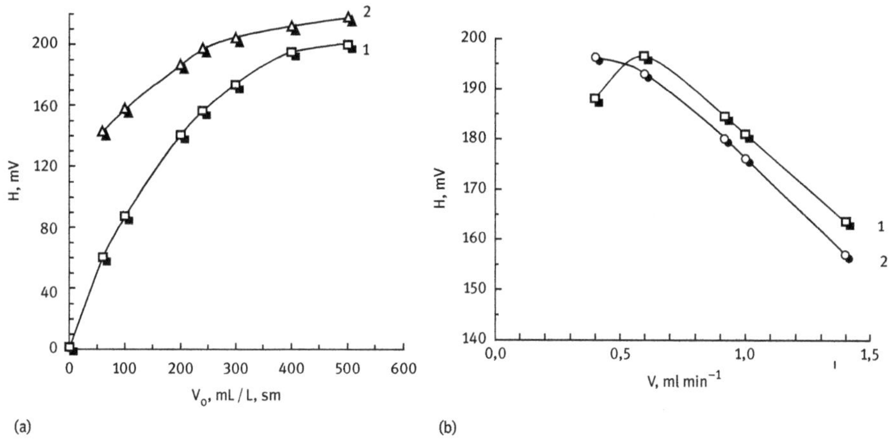

Figure 5.20: (a) Effect of different volumes (1) of 1 mmol/L Asc and different lengths of the reactor (2) to the signal; (b) effect of different flow rates of the sample solution (1) and probe molecule solution (2) to the signal.

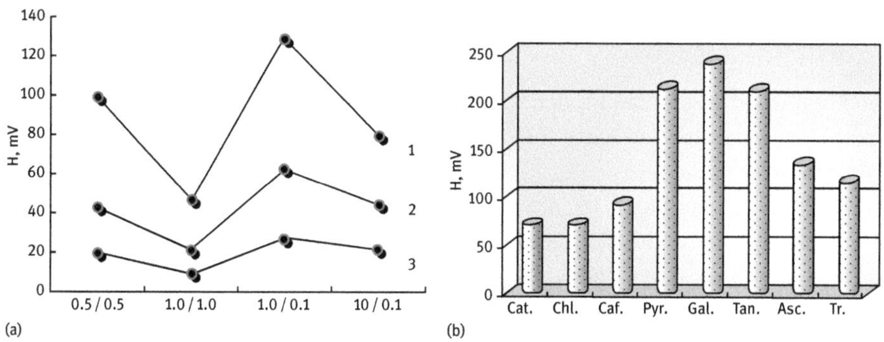

Figure 5.21: (a) Different ratios of $[Fe(CN)_6]^{3-}/[Fe(CN)_6]^{4-}$ to 0.2 mmol/L of Pyr (1), Asc (2), and Cat (b) assay signal for various antioxidants at 1.0 mmol/L $[Fe(CN)_6]^{3-}$/0.5 mmol/L $[Fe(CN)_6]^{4-}$ and pH 7.

Flow injection potentiometry (FIP) showed stable baseline and high sensitivity for different concentrations of antioxidants. The sensitivity of the assay is determined by the molecular structure of the antioxidant and its content in the sample. Figure 5.21(b) shows the results of the measurement with different numbers of phenolic hydroxyl antioxidants with the same concentration showing different potential signals. At the same time, pH is another important signal-affecting factor. Figure 5.22 shows the effect of pH on signals having different numbers of phenolic hydroxyl antioxidants. The increasing tendency of each antioxidant signal with increasing pH is mainly due to the dissociation constants of antioxidants, leading to changes in their protonation degree with the change in pH. With the increase of pH, they can more easily dissociate and then oxidize.

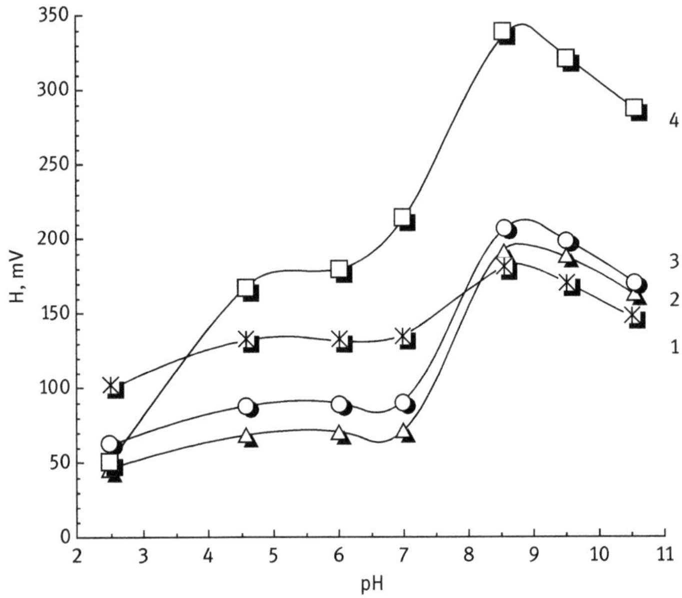

Figure 5.22: Effect of pH to the signal of Chl (1), Caf (2), Asc (3), and Pyr (4).

FIP is an ideal method for determining the antioxidant capacity of antioxidants, which can accurately measure the antioxidant capacity of various antioxidants at different pHs.

Under ideal conditions, ten antioxidants were measured and the results are listed in Table 5.1. Each antioxidant shows a good linear range within a certain range, where antithrombus, pyrogallol, and gallic acid show two linear range of segments. In addition, the linear range and sensitivity are affected by the ratio of potassium

Table 5.1: Linear equations of various antioxidants and related parameters (1.0 mmol/L $[Fe(CN)_6]^{3-}$/0.1 mmol/L $[Fe(CN)_6]^{4-}$ redox vs buffer solution).

Compound	Slope ($mVpC^{-1}$)	Correlation coefficient	Linear Range (mM)	LOD (μM)
Asc.	34.8	0.9975	0.01–0.10	9
	87.1	0.9998	0.10–1.00	
Cat.	64.4	0.9951	0.05–1.00	18
Tan.	114.6	0.9993	0.01–1.00	5
Pyr.	59.0	0.9941	0.01–0.10	8
	224.7	0.9977	0.10–1.00	
Chg.	45.8	0.9986	0.05–1.00	20

Table 5.1 (continued)

Compound	Slope ($mVpC^{-1}$)	Correlation coefficient	Linear Range (mM)	LOD (μM)
Caf.	67.1	0.9974	0.05–1.00	16
Gal.	59.6	0.9998	0.01–0.10	7
	262.4	0.9976	0.10–0.50	
Ur.	28.2	0.9969	0.10–1.00	25
Cys.	18.4	0.9972	0.10–1.00	32
Tr.	22.9	0.9987	0.05–1.00	30

ferrocyanide and potassium ferrocyanide in the probe molecule (Figure 5.23). The FIP recovery ranged from 98% to 102% and the standard deviation ranged from 0.7% to 1.0% (n = 5), suggesting that FIP is an effective method for determining the antioxidant capacity of antioxidants.

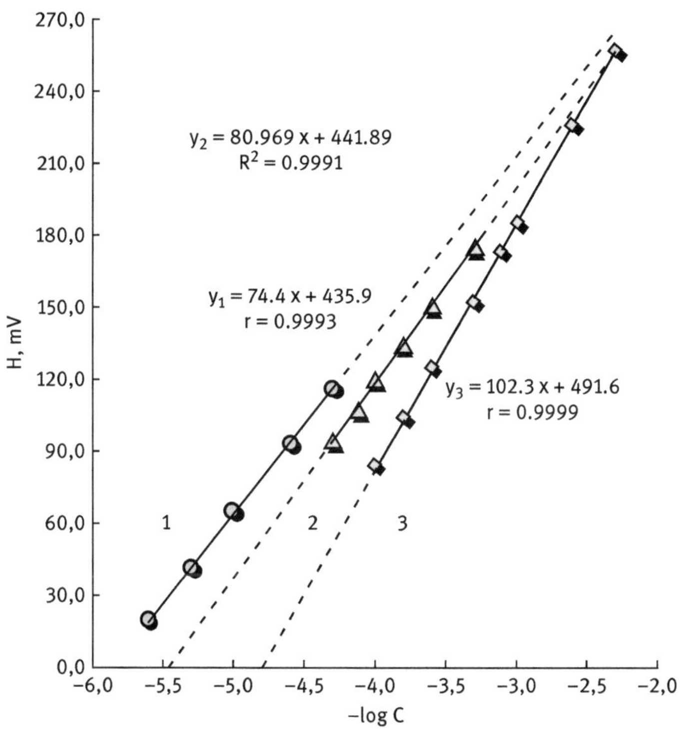

Figure 5.23: Influences of different ratios of $[Fe(CN)_6]^{3-}/[Fe(CN)_6]^{4-}$ (1–0.5 mmol/L/ 0.5 m, 2–1.0 mmol/L/0.5 m and 3–10.0 mmol/L/0.5 m) on Asc standard curve at pH = 8.5.

Because the antioxidative ability of antioxidants is affected by the concentrations of antioxidants, pH, and reaction substrates, the determination of the absolute value of individual antioxidants does not have a good persuasion, while the relative antioxidant activity under the same conditions (RAA_x) is of more significance. The relative antioxidant capacity (RAA_x) is given as follows:

$$RAA_X(\%) = \left(AA_{ref}/AA_x\right) \times 100$$

where AA_x is the antioxidant capacity of the test sample and AA_{ref} is the antioxidant capacity of the reference material.

The relative antioxidant capacity (RAA_x) of seven antioxidants was determined by antithrombus antioxidant as the reference. The results are listed in Table 5.2. The results of ascorbic acid, gallic acid, caffeic acid, and chlorogenic acid were compared with those obtained from the literature. Although the direct value does not fit with literatures, gallic acid among the antioxidants showed the highest antioxidant capacity in different techniques. Moreover, the value order that reflects the relative antioxidant capacity for these seven antioxidants is comparable with the referred literatures. The different results obtained by different methods are mainly due to different pH and concentrations at the time of determination. However, FIP shows good reproducibility when compared to that obtained by the TAS method.

Table 5.2: Results of different antioxidant capacity.

Compound	FIP method (Asc. as a ref.) (n =5, P = 0.95)		TAS	DPPH $[Tr.]_{eq}$ (mM)[b]	ABTS	AOP (mol/mol)[c]
	RAA (%)	RSD (%)				
Asc.	100	–	0.62	–	0.8	–
Cat.	54±1	1.5	–	1.9	–	–
Pyr.	227±3	1.0	–	–	–	–
Caf.	93±2	1.7	1.78	–	1.10,1.26,0.99	–
Chl.	91±2	1.8	–	1.9	–	38
Gal.	250±3	1.0	3.02	2.3, 4.0	2.45,2.69,3.01	2684
Tan.	435±4	0.7	–	–	–	–

Ascorbate antioxidant was the standard reference; the researchers measured the actual sample, as listed in Table 5.3. Green tea extract was 12.5–18.0 mmol/L, while black tea extract was 3.6–7.6 mmol/L, which is consistent with the results obtained by the literature, reporting that green tea showed stronger antioxidant capacity than black tea. Table 5.4 lists the antioxidant capacity of six Chinese medicine. As the traditional Chinese medicine contains a lower antioxidant, we use the standard addition

Table 5.3: Determination of total antioxidant capacity of tea extract ($n = 5$, $P = 0.95$).

	TAA (Asc.) eq (mM)	RSD (%)
	Black tea extracts	
Lipton yellow label tea	7.2 ± 0.1	1.1
Dolche vita	3.6 ± 0.1	2.2
Dilmach	4.5 ± 0.1	1.8
Riston	5.0 ± 0.1	1.6
Chelton	6.7 ± 0.1	1.2
May royal safari	3.8 ± 0.1	2.1
Greenfield golden ceylon	7.6 ± 0.1	1.0
	Green tea extracts	
Nadin	14.3 ± 0.2	1.1
Elite Chinese mint	12.5 ± 0.2	1.3
Elite lazure	15.4 ± 0.3	1.5
Green tea and chamomile	15.4 ± 0.3	1.5
May emerald valley	16.6 ± 0.3	1.4
Sweet osman	18.0 ± 0.2	0.9
Greenfield jasmine dream	16.9 ± 0.2	0.9

Table 5.4: Determination of the global antioxidant capacity of Chinese medicine extract.

Herbal Infusion	**TAA $[Asc.]_{eq}$ (mM)**			**RSD (%) ($n = 5$, $P = 0.95$)**
	Added (mM)	**Found (mM)**	**Recovery (%)**	
Lemon and ginger	–	0.515 ± 0.006	–	0.9
	0.50	1.035 ± 0.008	102.0	0.6
	1.00	1.994 ± 0.009	99.0	0.4
Pure peppermint	–	0.625 ± 0.004		0.5
	0.20	0.836 ± 0.007	101.3	0.7
Pure Howers of Camomile	–	0.920 ± 0.008		0.7
	2.00	2.934 ± 0.009	100.5	0.2
Camomiles and spearmint	–	2.41 ± 0.01		0.3
	0.60	2.95 ± 0.02	98.0	0.5
	2.00	5.10 ± 0.02	102.0	0.3
Mate		2.15 ± 0.01	–	0.4
Ginseng	–	2.07 ± 0.01	–	0.4

method. The recoveries ranged from 98.0% to 102.0%, and FIP showed low relative standard deviations (RSDs) of 0.2% to 0.9%. Finally, the eight kinds of fluids extracted from fruits were studied; the results are listed in Table 5.5. Antioxidant capacity of fruit extracts was from high to low for orange, grapefruit, pink grapefruit, kiwi fruit, apple, red grapes, and lemons. Measurement results are consistent with those obtained from the literature. The antioxidant order of various fruits is also affected by the pH and the quality of the fruits.

Table 5.5: Determination of global antioxidant capacity of fruit juice extract.

Fruit extract	TAA $[Asc.]_{eq}$ (mM)	RSD (%) ($n = 5$, $P = 0.95$)
Mandarin	0.370 ± 0.003	0.6
Orange	0.490 ± 0.001	0.2
Pink grapefruit	0.180 ± 0.001	0.4
Lemon	0.066 ± 0.002	2.4
Pomelo	0.280 ± 0.002	0.6
Red grape	0.130 ± 0.002	1.2
Kiwi	0.160 ± 0.003	1.5
Apple	0.148 ± 0.002	1.0

FIP is a convenient, highly sensitive, quick, and inexpensive antioxidant capacity measurement method. It has the following advantages: (1) a stable baseline measure and (2) can be used for a wide range of concentration. Given that this method has many advantages, it can be used for in vitro and also for the determination of antioxidant capacity of antioxidant in food and drink.

5.3.2 Voltammetric determination of antioxidant capacity of antioxidant

Research shows that phenolic substances act as antioxidants in the body and animal model, where phenol can retard atherosclerosis, thus reducing the risk of coronary artery disease, thereby minimizing the rate of mortality. Therefore, the determination of antioxidant capacity of phenolic antioxidants has a very vital significance. However, the current methods mostly use free radicals to determine phenolic antioxidants of biological samples, foods, extracts, and pure substances by vigorous reaction. There are few mild oxidation methods for determining the antioxidant capacity of phenolic antioxidants. Phenolic substances are generally less reductive substances, and their oxidation peaks can be clearly seen when measured by CV on a glassy carbon electrode. Therefore, the content of antioxidants can be determined by voltammetry based on their redox properties. Voltammetry showed that the bisphenol phenolic group had a lower oxidation potential than the ortho position and the monophenolic hydroxyl group. These results are consistent with the theoretical studies. Here voltammetry was used for the determination of antioxidant capacity of phenolic antioxidants in red wine and white wine.

First, the redox properties of standard phenolic antioxidants were studied. The results of CV for phenols, ascorbic acid, and sodium sulfate in simulated wines at three concentration gradients (0.01, 0.05, and 0.5 mmol/L) are listed in Table 5.6. The reaction of phenols follows the following formula:

$$\mathrm{R} \leftrightarrow \mathrm{O} + 2H^{+} + 2e^{-}$$

where R is an antioxidant (reducing agent) and O is the oxidation product.

Table 5.6 lists the measured results of the various electrochemical voltammetric methods for antioxidants. The peak potential was measured with good reproducibility, with a standard deviation of less than 3 mV. The anodic peak current ($I_{p,a}$) is proportional to the concentration of antioxidant, but the RSD is as high as 10% because of the need to reprocess the electrode surface for each measurement. The results of the electrochemical oxidation of diluted liquor and the first oxidation peak of 0.5, 0.05, 0.01 mmol/L antioxidant and the related parameter values are listed in Table 5.6 (d = 3 mm for a glassy carbon electrode, V = 100 mV/s, 12% ethanol solution, pH = 3.6 ± 0.2).

As can be seen from Table 5.6, when compared to other antioxidants, ascorbate begins to preferentially oxidize at 170 mV. The phenolic acids and flavonoids $E^{o\prime}$ containing orthophenolic hydroxyl groups were between 360 and 450 mV. After excluding the effect of pH, these results are similar to those obtained from the literature. Although morin does not have orthophenolic hydroxyl groups, it has a lower $E^{o\prime}$ (380 mV) and can be related to the similar structure of quercetin. In both molecules, the hydroxyl group replaces the C-3 position on the C-ring of flavonol, rendering the molecule a unique electrochemical behavior (more susceptible to oxidation) due to the electron-donating capability for ketone in the C-4 position and oxygen in the ring. However, only a phenolic hydroxyl is oxidized; the redox of morin is irreversible. Rutin has a rutinosaccharide linked to the C-3 position, which eliminates irreversibility of electrochemical redox reversion of rutin. However, such a structure also confers a higher oxidation potential of rutin than that of morin.

For diffusion-controlled reversible electrochemical reactions, the ideal value should be ($E_{p,a} - E_{p/2}$) = 28 mV at 25 °C. However, Table 5.6 shows that $|E_{p,a} - E_{p/2}|$ = 73 mV when the concentration of catechin is 0.5 mmol/L, indicating that catechins have lower electrochemical reversibility. The peak current difference ΔEp of catechin is 142 mV, which is far from the theoretical value of 29 mV. Thus, catechin shows its electrochemical irreversibility. However, at low concentrations, $|E_{p,a} - E_{p/2}|$ and ΔE_p of the antioxidants have lower values, indicating that the electrochemical reversibility of the antioxidants increases with a decrease in concentrations. The smaller values of

Table 5.6: Electrochemical characterization of the diluted wine solution and the first oxidation peak of 0.5, 0.05, and 0.01 mM antioxidant (d = 3 mm glassy carbon electrode, V = 100 mV•s^{-1}, 12% ethanol solution, pH = 3.6 ±0.2).

E in mV (Ag/AgCl)	Conc. (mM)	$E_{p,a}$	ΔE_p	$E_{p,a} - E_{p/2}$	E_{mid}	$(E_{p,a} + E_{p/2})/2$	$I_{p,a}$ (µA)	Q (µC) to 500 mV
				White Wines				
1999 Sauv. Blanc	(10 x dil.)	408	43	54	386	381	1.96±0.08	3.12±0.17
1984 Sauv. Blanc	(10 x dil.)	404	51	49	379	380	1.67 ± 0.10	1.92 ± 0.19
				Red Wines				
1998 Pinot Noir	(400 x dil.)	396	68	58	362	367	2.22 ± 0.14	3.40 ± 0.18
1991 Cab. Sauv.	(400 x dil.)	393	66	55	360	366	1.53 ± 0.04	2.20 ± 0.18
Ascorbic acid	0.5	242	[a]	66		209	12.0	21.7[b]
	0.05	215		53		188	1.24	2.3[b]
	0.01	211		50		186	0.25	0.49[b]
Quercetin	0.5[c]	376	28	43	362	355	3.1	3.2[d]
	0.05[c]	378	18	37	369	360	2.3	1.9[d]
	0.01	377	13	29	370	362	0.98	0.67[d]
Epicatechin	0.5	415	101	55	365	388	11.2	14.0
	0.05	380	45	37	357	361	4.3	4.1
	0.01	372	21	31	362	357	2.0	1.7

Catechin	0.5	455	142	73	384	419	10.7	12.7
	0.05	389	40	38	369	370	3.4	3.3
	0.01	379	19	29	369	364	1.18	0.91
Gallic acid	0.5	418		47		395	23.0	26.1
	0.05	391		32		375	3.3	3.5
	0.01	386		31		370	0.74	0.95
Morin	0.5	419		58		390	11.6	15.2
	0.05	398		44		376	3.9	4.8
	0.01	396		31		380	2.4	1.9
Tannic acid	0.5	429		53		403	9.7	11.4
	0.05	416		42		395	6.6	7.0
	0.01	405		36		387	1.45	1.43
Caffeic acid	0.5	435	100	52	385	409	20.7	22.2
	0.05	416	31	37	400	397	3.6	3.1
	0.01	410	17	35	401	392	1.46	1.6
Rutin	0.5	469	78	48	430	445	15.4	12.5
	0.05	471	33	37	454	452	4.3	3.0
	0.01	470	19	33	460	453	2.7	1.6
Ferulic acid	0.5	639		100		598	12.4	
Malvin	0.5	677		130		612	10.0	

(continued)

Table 5.6 (continued)

E in mV (Ag/AgCl)	Conc. (mM)	$E_{p,a}$	ΔE_p	$E_{p,a} - E_{p/2}$	E_{mid}	$(E_{p,a} + E_{p/2})/2$	$I_{p,a}$ (μA)	*Q* (μC) to 500 mV
t-Resveratrol	0.5[c]	667		79		627	3.1	
Vanillic acid	0.5	764		73		723	16.0	
p-Coumaric acid	0.5	804		112		748	13.9	
$Na_2S_2O_5$	0.5	970		190		875	18.5	

[a]Irreversible cyclic voltammetry test with no reduction peak.
[b]400 mV charging
[c]Partially dissolved
[d]450 mV charging

ΔE_p of quercetin are mainly due to the further oxidation of its oxidation products. However, ascorbic acid and gallic acid showed great irreversibility, as their oxidation products could not be reduced on the surface of a glassy carbon electrode.

At pH 7, the standard electrode potential for one electron oxidation for ascorbic acid is 282 mV (SHE). In this study, the obtained standard electrode potential is 189 mV (Ag/AgCl). Considering that the pK_a of ascorbic acid is 4.04 at pH 3.6, these two values are similar after relevant formula conversion. While ascorbic acid was introduced into the alcohol solution as antioxidant, different standard electrode potentials were obtained. Because ascorbic acid can be preferentially oxidized and protect other phenols from being oxidized, different forms of potential and lowest stability exhibited for ascorbic acid. Likewise, vitamin E performs similar nature.

Sodium thiosulfate showed a broader oxidation peak, with oxidation starting at 400 mV, but no peak was seen at 970 mV. Although the sulfite oxidation potential is very low (−420 mV (Ag/AgCl) at pH = 3.6) when calculated by thermodynamic free energy, in practice sulfite is more stable at the glassy carbon electrode than at low potentials; no oxidation current was observed mainly because of the energy barrier at the glassy carbon electrode during its reaction, which is very similar to the sulfur-containing compounds. When a very small amount of acetaldehyde is added to the reaction solution, all sulfite is eliminated so that the oxidation peak of the solution disappears, indicating that only free sulfite is electrochemically active. Therefore, sulfite is added to wine. Phenols can be protected from oxidation; at the same time, the electrochemical measurement does not affect the phenolic measurement signal.

Other phenols, such as ferulic acid, tertiary resveratrol, mevalonide, vanillic acid, and p-coumaric acid, exhibited higher potentials (Table 5.6) due to the lack of ortho-diphenolic hydroxyl groups and these phenolic compounds have a lower antioxidant capacity. Oxidation of these phenols involves one or two electrons, exhibiting a broader oxidation peak and lower reversibility. Some phenols showed lower reduction peaks, and vanillic acid showed lower than predicted reduction peaks due to the reduction of various oxidation products. However, some of the oxidation products irreversibly adsorbed to the electrode surface. In this case, the electrode must be reprocessed to renew the electrode surface for the next use.

There are several phenols with the first oxidation peak between 370 and 470 mV and the second oxidation peak (<800 mV) at a higher potential, such as vanillic acid. The same phenomenon can be observed with catechin, epicatechin, quercetin, and gallic acid. The resultant electrode contamination by oxidation products also increases the technical difficulty of voltammetric analysis of antioxidants.

Catechins and caffeic acid were analyzed with simulated diluted wine and the determined current intensity was found to be proportional to the concentration of antioxidants. Although the peak currents of catechins and caffeic acid differ significantly at high concentrations (0.5 mmol/L), the currents were similar and linear at low concentrations (<2 μmol/L; Figure 5.24 (b)). Ascorbic acid was linear in the range 0.01–0.5 mmol/L. Other antioxidants did not exhibit a linear relationship between peak

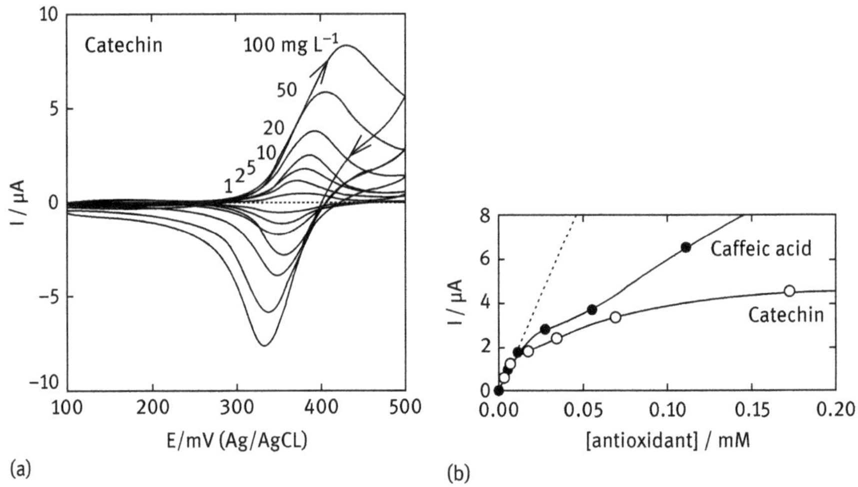

Figure 5.24: (a) Cyclic voltammograms of catechins at various concentrations (background subtracted); and (b) peak current (catechin and caffeic acid) versus concentration.

current and concentration at high concentration. The $I_{p,a}/I_{p,c}$ of catechin and epicatechin were approximately 0.7 at high concentrations (0.5 mmol/L), while it was gradually close to 1 with dilution of both of them. It is indicated that the lower is the antioxidant concentrations, the more reversible is the electrochemical behavior. This phenomenon is also observed for caffeic acid and rutin.

Except quercetin and tannic acid, the first oxidation peak currents of six phenolic compounds were 3.8 (±0.4) µA and 1.7 (±0.7) µA at 0.05 and 0.01 mmol/L, respectively. Different antioxidants with different peak currents are mainly due to different diffusion coefficients and electrode surface energy barrier; these differences increase owing to electrode surface oxide adsorption pollution. Gallic acid had the highest peak current at 0.5 mmol/L and the lowest peak current at 0.01 mmol/L, indicating that gallic acid had the lowest absolute current at the glassy carbon electrode, but more effective peak current at higher concentration. Similar to the F–C optical method, the standard curve of gallic acid is often used as a standard equivalent to determine the amount of antioxidant (e.g., wine) in a mixture.

As listed in Table 5.6, the amount of charge flowing through the antioxidant when voltammetrically scanned on a glassy carbon electrode was also recorded. The highest potentials for ascorbic acid and quercetin scan were 400 and 450 mV, while the other antioxidant scan had the highest potential of 500 mV. The calculation of the antioxidative capacity of antioxidants by measuring the charge flow through the other antioxidants in the wine as the standard equivalent of the amount of charge flowing through gallic acid shows advantages over the peak current because many antioxidants in mixtures (wine) have similar peak potentials, thereby the inaccuracies in peak current calculations.

A similar method can also be used to determine the antioxidant capacity of antioxidants in white wines. Antioxidant species in white wine are gradually oxidized as the potential applied to the glassy carbon electrode approaches its standard electrode potential. When the antioxidant on the electrode surface is depleted, the current will reach a peak value and then decrease with diffusion of the bulk antioxidant. With less antioxidant species, it is possible to produce only a small shoulder on the main current peak.

Figure 5.25(a) shows the cyclic voltammogram of a 1999 Sauvignon Blanc white wine. It can be seen from the figure that the blank simulated wine did not show obvious oxidation peak before 1000 mV because the oxidation of ethanol was higher than 1000 mV. The sample wine has two peaks at 400 and 900 mV, with peak currents of 12 and 25 μA, respectively. At currents of 200–500 mV, the peaks are mainly caused by phenols with lower standard electrode potentials and added ascorbic acid. The first peak current can quantify the amount of antioxidants it contributes, but the dilution of the wine must be within the linear range of the standard. Figure 5.26(a) shows that the liquor is diluted tenfold to meet the requirement. The first peak current is about 2 μA after tenfold dilution, and the RSD of the electrode is about 5%. If the electrode surface undiluted most of the antioxidant oxidation of wine, the current value can reach 20 μA; thus, the reverse scan reduction current is submerged. Although undiluted wine samples showed irreversibility of antioxidants, the reduction peak gradually appeared with the dilution of wine samples (Figure 5.26 (a)), indicating that the oxidation–reduction characteristics of antioxidants in high-fold diluted wine samples were close to quasireversible. The ratio of $I_{p,a}/I_{p,c}$ increased from 0.30 diluted 10 times to 0.49 diluted 50 times, while ΔE_p decreased from 57 mV diluted 5 times to 22 mV diluted 50 times. With the increase in dilution multiples, its redox properties are closer to reversible.

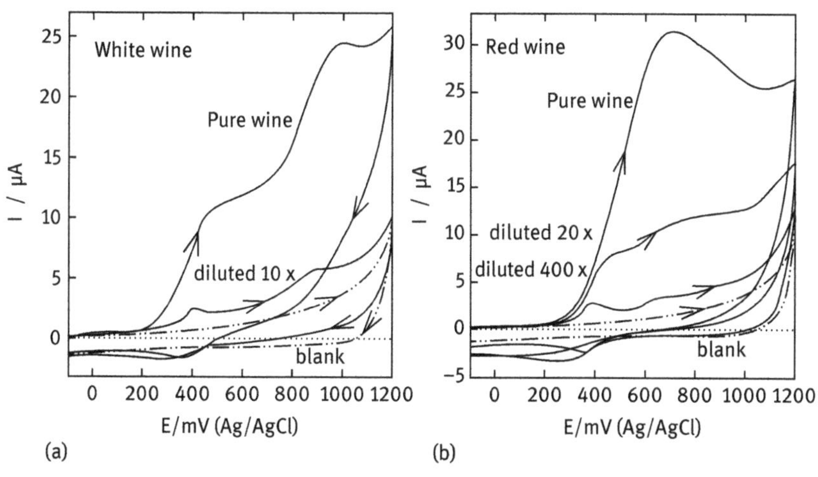

Figure 5.25: Cyclic voltammograms of different dilutions of white wine (a) and red wine (b) (12% ethanol solvent, pH = 3.6, sweep rate 100 mV •s^{-1}).

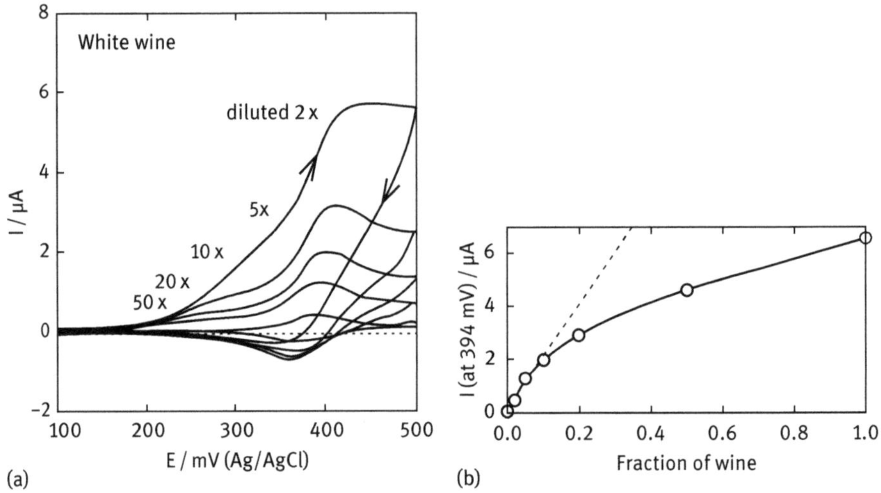

Figure 5.26: (a) Cyclic voltammograms of white wine with different dilution times (background subtracted). (b) Curves of oxidation current at 394 mV versus different percentages of alcohol.

The peak current of tenfold diluted wine was 1.98 (±0.08) µA, equivalent to 52 mg/L gallic acid signal. However, compared with one certain kind of phenol, the 10-fold dilution system showed a wide potential difference (| E p, a -E p/2 | = 54 mV), which should be ascribed to many different phenols with similar standard electrode potential that coexists in white wines. If the result is scaled to 500 mV, the diluted white wine has a charge of 3.12 (0.17) µC, equivalent to 78 mg/L of gallic acid. From the F–C method, the result is equivalent to 322 mg/L of gallic acid, which is 4–5 times the CV. This is mainly because many phenols at 500 mV cannot still be electrode oxidized, and thus these do not contribute to the measured results.

The 1999 Sauvignon Blanc white wine diluted 10 times and this showed a significant second oxidation peak at 888 mV. After subtracting the background current, the oxidation peak becomes more pronounced (Figure 5.27(a)). Figure 5.27(b) shows a relationship between the accumulated charge and potential. The oxidation current of 700–1000 mV is mainly composed of vanillic acid, the oxidation of coumaric acid, and the secondary oxidation of flavonoids. The standard electrode potential of nearly 600 mV of antioxidants, such as ferulic acid oxidation peak current, is mainly between the first oxidation peak and the second oxidation peak. This is due to the fact that the oxidation peak is weak and cannot show significant reversible oxidation peak.

The 1999 Sauvignon Blanc cyclic voltammograms of white wine shows a typical graph, but other wines have some differences. When the second wine was selected as the analyte, the detection data were shown in Figure 5.27(a). It is found that the first oxidation peak was more strident, and the peak current and charge

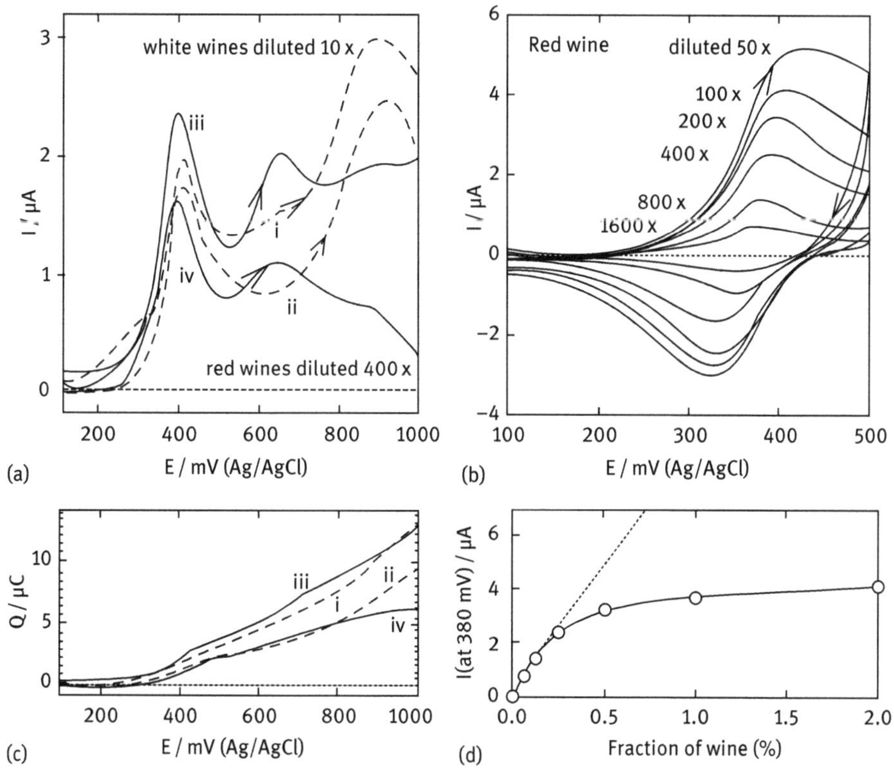

Figure 5.27: (a) Ten times diluted white wine (i) 1999 Sauvignon Blanc, (ii) 1984 Sauvignon Blanc, diluted 400 times red wine, (iii) 1998 Pinot Noir, (iv) 1991 Cabernet Sauvignon linear scan; (b) relationship between charge and potential of white wine and red wine; (c) cyclic voltammograms of different dilutions of red wine (net background); and (d) 380 mV oxidation current versus different percentages of wine.

(scan to 500 mV) were 44 and 41 mg/L of gallic acid equivalent. Further investigation revealed that this wine shows weaker antioxidant capacity than that of 1999 Sauvignon Blanc white wine. The 1984 Sauvignon Blanc white wine showed weaker antioxidant capacity. When the F–C method is used to test, the determination result is equivalent to 269 mg/L of gallic acid.

Figure 5.25(b) shows the redox peaks of the 1998 Pinot Noir red wine. With pure red wine, only a large oxidation peak can be observed. Red wine samples need to be diluted 400 times within the linear range, indicating that red wine has a higher antioxidant capacity. ΔE decreases with increasing dilution (28 mV at 1600 dilution), showing good quasireversibility (Figure 5.27(c)).These results were also validated by the ratio of $I_{p,a}/I_{p,c}$, with a ratio of 0.56 at 50-fold dilution and 0.71 at 1600-fold dilution. When compared with white wine, red wine shows a stronger reduction current, mainly because of the fact that red wine contains a higher proportion of redox reversible antioxidants, such as catechins and caffeic acid. On the other hand, the oxidation current of red wine reached the upper limit of its linear range by 2 μA, which was more linear with increasing concentration (Figure 5.27(d)), indicating that the complex components of red wine more easily contaminated the electrode surface.

For undiluted red wine, the oxidation peak current was 2.22 (±0.14) μA, equivalent to 2400 mg/L of gallic acid. Using electricity law, it can be expressed as 3.40 (±0.18) μC, equivalent to 3580 mg/L of gallic acid, which in turn means equivalent to 40 times of the white wine. For 1998 Pinot Noir red wine, the results obtained by electrochemical methods and F–C determination method can be compared. However, for the 1991 Cabernet Sauvignon, the peak current and the scan to 500 mV were equivalent to 1605 and 1945 mg/L of gallic acid, respectively, whereas the F–C assay was equivalent to 3342 mg/L of gallic acid. It is important to emphasize that CV and F–C methods respond to different phenols accordingly, whereas electrochemical coulometry can be used only for the most reactive antioxidants.

Unlike white wine, the oxidation peak at 651 mV of red wine, minus the background current, became more pronounced (Figure 5.27(a)), which can be attributed to delignin and anthocyanin contents in red wine. This is further evidenced by the studying the aged 1991 wine, which shows that the oxidation peak at 651 mV is relatively weaker because of its reduced anthocyanin content with an increase in the storage time. At the same time, ferulic acid and tertiary resveratrol also contributed to the oxidation peak current. However, the oxidation peak current at 893 mV is less pronounced because of the lower phenolic content of the high standard electrode potential.

CV can be used to measure the antioxidant capacity of standard antioxidants and white and red wine effectively. This method is economical, inexpensive, and easy to handle. CV also measures the antioxidant capacity of biological and food extracts at various pH levels. CV of antioxidants in wine can help determine the antioxidant capacity, which is conducive to the development of wine nutrition standards. The method can not only monitor the degree of aging in wine but can also help monitor the wine industry.

5.3.3 Antioxidant capacity of antioxidants measured by the DNA damage method

During normal physiological metabolism, some substances with high oxidation activity (ROS and RNS), such as superoxide radical, nitrogen radical, hydroxyl radical, nitric oxide, and hydrogen peroxide, are produced in the body. Normal concentrations of ROS and RNS are good for the health of the organism, but once the concentrations of ROS and RNS are high, hydroxyl radicals, in particular, will severely damage the biological macromolecules of DNA, proteins, and lipids, causing pathological changes in cells and tissues, leading to cancer, cardiovascular disease, diabetes, neurodegeneration, and other diseases. Glutathione, lipid acids, uric acid, flavonoids, and other antioxidants can effectively eliminate the ROS and RNS to keep the body healthy, and so antioxidant capacity evaluation is of great significance. DNA is the main target for oxidative damage of hydroxyl radicals in the human body. Therefore, the determination of the antioxidant capacity using DNA as a target molecule to the largest content simulate the real situation of endogenous free radical active species in the human body. Currently, free radicals widely used in various methods for measuring antioxidant capacity are synthetic free radicals such as 1,1-diphenyl-2-picrylhydrazyl radical 2,2-Diphenyl-1-(2,4,6-trinitrophenyl) hydrazyl (DPPH), 2,2′-azinobis-(3-ethylbenzthiazoline-6-sulfonate) (ABTS), and galvinoxyl. However, there are some differences between synthetic free radicals and ROS in vivo, and it is difficult to simulate the real situation in a human body with the results obtained from determination of antioxidant capacity. Hydroxyl radicals have better ability to reflect the actual situation in vivo and high activity for the determination of antioxidant capacity in vitro. Currently there are two methods used for the generation of hydroxyl radicals: one is the chemical method, such as Fenton reaction and the other is photocatalysis. When compared to the chemical method, the optical method has drawn the more attention because of its simple and controllable operation.

Since 1972 when Professor Fujishima and Honda published a report that titanium dioxide can generate hydrogen and oxygen under light, titanium dioxide has attracted a great deal of interest from photocatalytic scientists owing to its easy preparation, stability, and nontoxic properties. However, its wide bandgap (3.2 eV) makes it capable of absorbing only UV light; therefore it cannot use a large proportion of visible light, which limits its development and application as a photocatalyst. At the same time, UV light can damage the DNA, which interferes with the determination of the oxidation resistance of the probe where DNA acts as a molecular probe. Therefore, it is of great significance to develop a photocatalytic material capable of generating hydroxyl radicals in visible light to measure the oxidation resistance.

In vivo high concentrations of hydroxyl radicals can damage DNA and can cause diseases, while antioxidants compete with DNA with hydroxyl radicals to protect DNA from the damage. Because it can well simulate the real situation of antioxidation in vivo, DNA as a probe for the evaluation of antioxidant capacity has drawn considerable attention of researchers.

Here, a TiO_2 complex is taken as a photocatalytic hydroxyl radical source, and DNA damage is evaluated. Figure 5.28 shows the agarose gel electrophoresis of DNA. The natural DNA electrophoresis showed that the fragment was more than 2000 bp (band e), but most of the DNA fragments were less than 1000 bp (band b) after 30 min of irradiation, indicating that the DNA was damaged. When the detailed concentration of gallic acid is 30 mmol/L, the DNA fragment results in about 2500 bp (b and d) under the same irradiation conditions, indicating that the antioxidant, gallic acid, can effectively protect the DNA from hydroxyl radical damage.

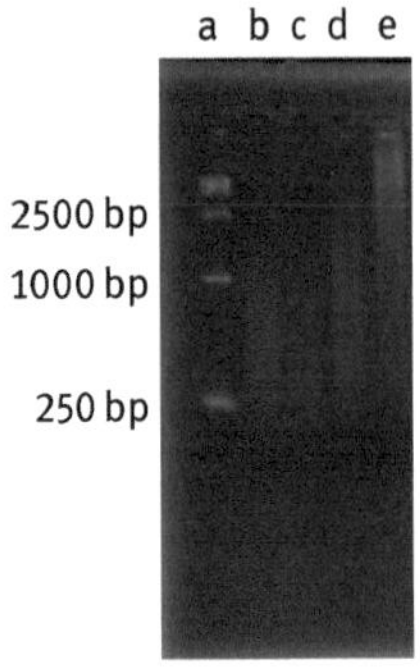

Figure 5.28: DNA gels processed under various conditions. Band a: DNA labeling; band b and c: DNA was irradiated for 20 and 30 min, respectively, in the presence of the TiO_2 complex; band d: addition of gallic acid, with the rest of the conditions as in c; band e: DNA was irradiated without TiO_2 complex for 30 min.

Because of its high sensitivity and convenience, electrochemical methods have also been used to determine DNA damage. Of the four bases that consist of DNA, the G-base oxidation potential is the lowest, which is often applied to reflect the extent of DNA damage by measuring its electrochemical signal.

As shown in Figure 5.29(a), the G base shows an oxidation peak at 0.92 V. With an increase in the irradiation time, the oxidation peak of G base decreases gradually, indicating that DNA has been oxidatively damaged by hydroxyl radicals. From the discussion, we can see that the results of electrochemical measurements and those of agarose gel electrophoresis are consistent.

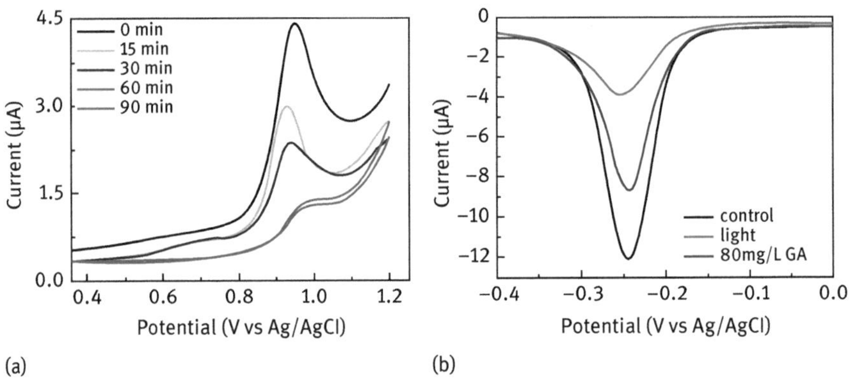

Figure 5.29: (a) Electrochemical LSV patterns of G bases at different irradiation times; (b) electrochemical differential pulse voltammetry patterns of MB-labeled DNA after 80 min of exposure to light and illumination with addition of 80 mg/L GA.

These studies have shown that TiO_2-based photocatalytic complexes can serve as a source of hydroxyl radicals that mimic endogenous free radicals, and their DNA damage can be used for the assessment of antioxidant capacity of antioxidant.

There have been many related studies on the use of DNA and G bases as molecular probes for the detection of antioxidants. However, because of the higher oxidation potential of DNA, direct electrochemical detection of DNA is often interrupted by reducing substances in the detection solution. MB molecules are often used to label DNA for electrochemical detection of DNA due to its lower redox potential and ease of binding to double-stranded DNA.

As shown in Figure 5.29(b), in the absence of light, the double-stranded DNA binds to a large number of MB molecules and shows a strong electrochemical signal. After the DNA is irradiated with light for 30 min, the DNA molecules are damaged, thereby decreasing the number of MB-binding molecules and the electrochemical signal decreases as well. When 80 mg/L GA was added, the DNA was irradiated again, because the DNA in the presence of GA was protected and the electrochemical signal also increased. By adding different concentrations of GA, electrochemical signals increased to varying degrees. On the basis of this principle, antioxidant capacity of antioxidants can be determined.

Many food items contain antioxidants, and so have a strong antioxidant capacity. Figure 5.30 shows the determination of the antioxidant capacity of GA. With the addition of different concentrations of GA, the reduction current of MB gradually increases, in the order of 10–100 mg/L. The linear equation is $y = 0.062x + 2.846$ ($R^2 = 0.991$); the limit of detection was 0.85 mg/L. When the concentration of GA was 20 mg/L, the RSD was only 3.2%. Compared with the previous method, this method also showed a wide linear range (0.10–0.50 mg/L). When the concentration of GA is 20 mg/L, it was found that 1,000 times of glucose, fructose, malic acid and glycine, 100 times of cysteine in concentration did not interfere with the determination results. The above investigations proved that this strategy is competent for antioxidant capacity assay in a complex system.

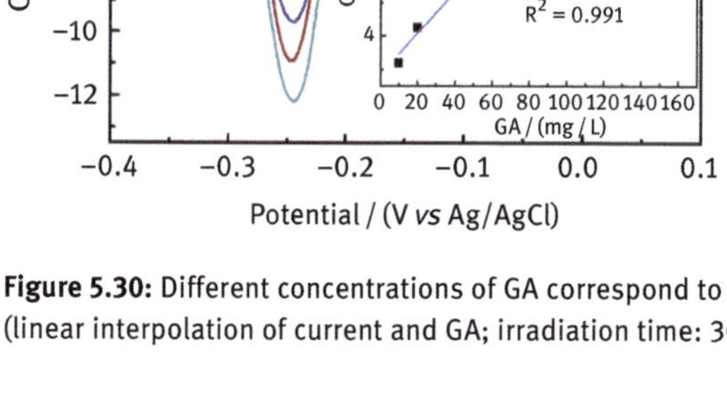

Figure 5.30: Different concentrations of GA correspond to different reduction currents of MB (linear interpolation of current and GA; irradiation time: 30 min).

To compare the antioxidant capacity of different antioxidants, researchers compared several antioxidants (ascorbic acid, glutathione, water-soluble vitamin E mimics (TROLOX), GA, and resveratrol RES). Figure 5.31 shows the antioxidant efficiency of different antioxidants. Antioxidant efficiency is defined as follows: % efficiency = $I_b/I_a \times 100\%$, where I_a is the not damaged MB reduction current, and I_b is the addition of antioxidants in the light. Ascorbic acid showed the highest antioxidant efficiency. This result was confirmed by the results of Barroso. Glutathione and TROLOX showed lower antioxidant efficiency, which was 17.0% and 20.0%, respectively. The antioxidant efficiencies of GA and RES were 27.5% and 30.1%, respectively.

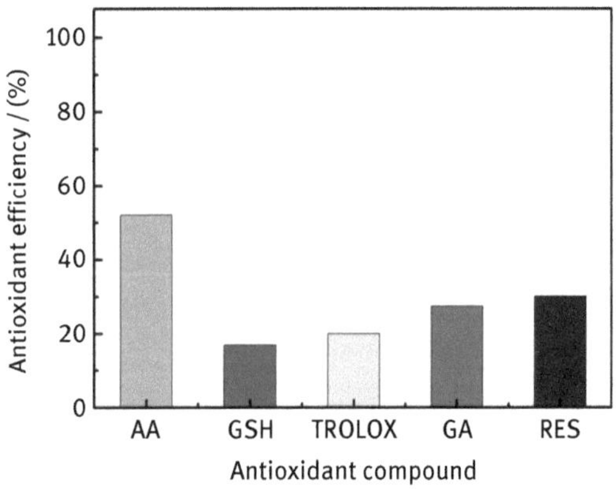

Figure 5.31: The hydroxyl radical eliminating abilities for ascorbic acid (AA), glutathione (GSH), water-soluble vitamin E mimics (TROLOX), GA, and resveratrol RES).

Tea contains polyphenols and some acids and has a strong antioxidant capacity. We use this method on five different brands of tea to test the antioxidant capacity; the results are listed in Table 5.7. As researchers expected, all tea samples show a strong antioxidant capacity. Among these, green tea (B1, B2, and B3) showed the strongest antioxidative capacity of 47.45, 66.61, and 41.86 mg/g, respectively. B4 and B5 showed relatively weak antioxidative capacity, which may be related to the process of tea production – the production process through the fermentation and drying process. In these processes, a part of the phenolic material is oxidized, resulting in lower antioxidant capacity.

To further confirm the reliability of this test, the researchers used two spectroscopic methods to determine the antioxidant capacity of tea leaves. The results of the F–C method showed the same trend as that of the DNA method. However, the results of the F–C method were generally higher than that of the DNA method. This is probably because the result obtained by the spectral F–C method is disturbed by the color of the tea sample. The results of another spectroscopic method (DPPH) also showed the same pattern as that of the DNA-sensing method. Therefore, the

Table 5.7: The test results of antioxidant capacity assay of five brands of tea using three different methods.

	DNA Sensor GA (mg•g^{-1})	F-C Method GA (mg•g^{-1})	DPPH Method Trolox (mg•g^{-1})
B1	47.45	60.29	141.20
B2	66.61	62.94	201.10
B3	41.86	78.40	191.50
B4	24.34	18.55	49.90
B5	38.39	39.82	86.30

DNA-sensing method based on photoelectrocatalytic can be used to detect the antioxidant capacity of antioxidants in tea.

By simple synthesis of photocatalytic material, the generation of endogenous free radicals can be effectively simulated, thus damaging DNA. Hence, the electrochemical signal of the DNA marker molecule MB is reduced. After an antioxidant is added, it can competitively react with DNA to protect the DNA, thereby reducing its damage and increasing the electrochemical signal of the labeled molecule MB. On the basis of this principle, the antioxidant capacity of antioxidants in tea was studied. When compared with the traditional method, this type of the sensing technique method has the following advantages: (i) Using hydroxyl radicals helps determine more effectively the in vivo antioxidant status; (ii) when compared with the chemical method (the Fenton reaction), the free radicals produced by the method are simple and can be easily controlled; and (iii) since DNA was selected as the target molecule, the measurement results can well simulate the real situation in the body. The DNA sensing method can be effectively used for the determination of antioxidant capacity of antioxidants in industrial food and biomedicine industries.

References

[1] Herrero-Martinez J M, Repolles C, Bosch E, et al. Potentiometric determination of aqueous dissociation constants of flavonols sparingly soluble in water. Talanta, 2008, 74: 1008–1013.

[2] Oliveira C M, Ferreira A C S, de Pinho P G, et al. Development of a potentiometric method to measure the resistance to oxidation of white wines and the antioxidant power of their constituents. Journal of Agricultural and Food Chemistry, 2002, 50: 2121–2124.

[3] Abdullin I F, Turova E N, Ziyatdinova G K, et al. Potentiometric determination of ascorbic acid: Estimation of itscontribution to the total antioxidant capacity of plant materials. Journal of Analytical Chemistry, 2002, 57: 353–355.

[4] Evtugyn G A, Shamagsumova R V, Stoikova E E, et al. Potentiometric sensors based on polyaniline and thiacalixarenes for green tea discrimination. Electroanalysis, 2011, 23: 1081–1088.

[5] Tessutti L S, Macedo D V, Kubota L T, et al. Measuring the antioxidant capacity of blood plasma using potentiometry. Analytical Biochemistry, 2013, 441: 109–114.
[6] Brainina K Z, Ivanova A V, Sharafutdinova E N, et al. Potentiometry as a method of antioxidant activity investigation. Talanta, 2007, 71: 13–18.
[7] Lee G J, Lee S K, Kim J M, et al. Application feasibility of antioxidant activity evaluation using potentiometry in major depressive disorder. Electrochemistry, 2014, 82: 264–266.
[8] Ruffien-Ciszak A, Baur J, Gros P, et al. Electrochemical microsensors for cutaneous surface analysis: Application to the determination of pH and the antioxidant properties of stratum corneum, Irbm. Irbm, 2008, 29: 162–170.
[9] Brainina K Z, Alyoshina L V, Gerasimova E L, et al. New electrochemical method of determining blood and blood fractions antioxidant activity. Electroanalysis, 2009, 21: 618–624.
[10] Brainina K Z, Gerasimova E L, Kasaikina O T, et al. Antioxidant activity evaluation assay based on peroxide radicals generation and potentiometric measurement. Analytical Letters, 2011, 44: 1405–1415.
[11] Nieh C H, Hsieh B C, Chen P C, et al. Potentiometric flow-injection estimation of tea fermentation degree. Sensors and Actuators B. 2009, 136: 541–545.
[12] Sano M, Yoshida R, Degawa M, et al. Determination of peroxyl radical scavenging activity of flavonoids and plant extracts using an automatic potentiometric titrator. Journal of Agricultural and Food Chemistry, 2003, 51: 2912–2916.
[13] Chevion S, Roberts M A, Chevion M. The use of cyclic voltammetry for the evaluation of antioxidant capacity. Free Radical Biology and Medicine, 2000, 28: 860–870.
[14] Kilmartin P A. Electrochemical detection of natural antioxidants: Principles and protocols. Antioxidantsand Redox Signaling, 2001, 3: 941–955.
[15] Zhang D, Chu L, Liu Y, et al. Analysis of the antioxidant capacities of flavonoids under different spectrophotometric assays using cyclic voltammetry and density functional theory. Journal of Agricultural and Food Chemistry, 2011, 59: 10277–10285.
[16] Roginsky V, Barsukova T, Hsu C F, et al. Chain-breaking antioxidant activity and cyclic voltammetry characterization of polyphenols in a range of green, oolong, and black Teas. Journal of Agricultural and Food Chemistry, 2003, 51: 5798–5802.
[17] Gil E S, Couto R O. Flavonoid electrochemistry: A review on the electroanalytical applications. Revista Brasileira De Farmacognosia-Brazilian Journal of Pharmacognosy, 2013, 23: 542–558.
[18] Kilmartin P A, Zou H L, Waterhouse A L. A cyclic voltammetry method suitable for characterizing antioxidant properties of wine and wine phenolics. Journal of Agricultural and Food Chemistry, 2001, 49: 1957–1965.
[19] Piljac-Zegarac J, Valek L, Stipcevic T, et al. Electrochemical determination of antioxidant capacity of fruit tea infusions. Food Chemistry, 2010, 121: 820–825.
[20] Masoum S, Behpour M, Azimi F, et al. Potentiality of chemometric approaches for the determination of (+)-catechin in green tea leaves at the surface of multiwalled carbon nanotube paste electrode. Sensors and Actuators B-Chemical, 2014, 193: 582–591.
[21] Glod B K, Kiersztyn I, Piszcz P. Total antioxidant potential assay with cyclic voltammetry and/or differential pulse voltammetry measurements. Journal of Electroanalytical Chemistry, 2014, 719: 24–29.
[22] Souza L P, Calegari F, Zarbin A J G, et al. Voltammetric determination of the antioxidant capacity in wine samples using a carbon nanotube modified electrode. Journal of Agricultural and Food Chemistry, 2011, 59: 7620–7625.
[23] Yardim Y, Keskin E, Senturk Z. Voltammetric determination of mixtures of caffeine and chlorogenic acid in beverage samples using a boron-doped diamond electrode. Talanta, 2013, 116: 1010–1017.

[24] Bordonaba J G, Terry L A. Electrochemical behaviour of polyphenol rich fruit juices using disposable screen-printed carbon electrodes: Towards a rapid sensor for antioxidant capacity and individual antioxidants. Talanta, 2012, 90: 38–45.
[25] Kakran M, Sahoo N G, Bao H, et al. Functionalized graphene oxide as nanocarrier for loading and delivery of ellagic acid. Current Medicinal Chemistry, 2011, 18: 4503–4512.
[26] Zhang Y, Liu Y, Yang Z, et al. Rapid electrochemical detection of ferulic acid based on a graphene modified glass carbon electrode. Analytical Methods, 2013, 5: 3834–3839.
[27] Curulli A, Di Carlo G, Ingo G M, et al. Chitosan stabilized gold nanoparticle-modified Au electrodes for the determination of polyphenol index in wines: A preliminary study. Electroanalysis, 2012, 24: 897–904.
[28] Sharpe E, Frasco T, Andreescu D, et al. Portable ceria nanoparticle-based assay for rapid detection of food antioxidants (NanoCerac). Analyst, 2013, 138: 249–262.
[29] Jakubec P, Bancirova M, Halouzka V, et al. Electrochemical sensing of total antioxidant capacity and polyphenol content in wine samples using amperometry online-coupled with microdialysis. Journal of Agricultural and Food Chemistry, 2012, 60: 7836–7843.
[30] Crevillen A G, Avila M, Pumera M, et al. Food analysis on microfluidic devices using ultrasensitive carbon nanotubes detectors. Analytical Chemistry, 2007, 79: 7408–7415.
[31] Crevillen A G, Pumera M, Cristina Gonzalez M, et al. Towards lab-on-a-chip approaches in real analytical domains based on microfluidic chips/electrochemical multi-walled carbon nanotube platforms. Lab on a Chip, 2009, 9: 346–353.
[32] Kovachev N, Canals A, Escarpa A. Fast and selective microfluidic chips for electrochemical antioxidant sensing in complex samples. Analytical Chemistry, 2010, 82: 2925–2931.
[33] Cortina-Puig M. Determination of the antioxidant capacity of dietary substances and extracts from natural sources at molecular, cellular and whole tissue level: Sensors and complex microfluidic structures. Current Analytical Chemistry, 2012, 8: 427–427.
[34] Amatatongchai M, Laosing S, Chailapakul O, et al. Simple flow injection for screening of total antioxidant capacity by amperometric detection of DPPH radical on carbon nanotube modified-glassy carbon electrode. Talanta, 2012, 97: 267–272.
[35] Mrazek N, Watla-iad K, Deachathai S, et al. Rapid antioxidant capacity screening in herbal extracts using a simple flow injection-spectrophotometric system. Food Chemistry, 2012, 132: 544–548.
[36] Oliveira R, Bento F, Sella C, et al. Direct electroanalytical method for alternative assessment of global antioxidant capacity using microchannel electrodes. Analytical Chemistry, 2013, 85: 9057–9063.
[37] Actis P, Tokar S, Clausmeyer J, et al. Electrochemical nanoprobes for single-cell analysis. Acs Nano, 2014, 8: 875–884.
[38] Lin Y, Yu P, Hao, J, et al. Continuous and simultaneous electrochemical measurements of glucose, lactate, and ascorbate in rat brain following brain. Analytical Chemistry, 2014, 86: 3895–3901.
[39] Xiang L, Yu P, Hao J, et al. Vertically aligned carbon nanotube-sheathed carbon fibers as pristine microelectrodes for selective monitoring of ascorbate in vivo. Analytical Chemistry, 2014, 86: 3909–3914.
[40] Liu K, Yu P, Lin Y, et al. Online electrochemical monitoring of dynamic change of hippocampal ascorbate: Toward a platform for in vivo evaluation of antioxidant neuroprotective efficiency against cerebral ischemia injury. Analytical Chemistry, 2013, 85: 9947–9954.
[41] Wei X, Liu C, Tu Y. Microemulsion-enhanced electrochemiluminescence of luminol-H2O2 for sensitive flow injection analysis of antioxidant compounds. Talanta, 2012, 94: 289–294.

[42] Cheng C, Huang Y, Wang J, et al. Anodic electrogenerated chemiluminescence behavior of graphite-like carbon nitride and its sensing for rutin. Analytical Chemistry, 2013, 85: 2601–2605.
[43] Goodney D E, Silverstein T P. Using the tyrosinase-based biosensor to determine the concentration of phenolics in wine. Journal of Chemical Education, 2013, 90: 1710–1712.
[44] Nadifiyine S, Haddam M, Mandli J, et al. Amperometric biosensor based on tyrosinase immobilized on to a carbon black paste electrode for phenol determination in olive oil. Analytical Letters, 2013, 46: 2705–2726.
[45] Qu Y, Ma M, Wang Z, et al. Sensitive amperometric biosensor for phenolic compounds based on graphene-silk peptide/tyrosinase composite nanointerface. Biosensors and Bioelectronics, 2013, 44: 85–88.
[46] Tang L, Zhou Y, Zeng G, et al. A tyrosinase biosensor based on ordered mesoporous carbon-Au/l-lysine/Au nanoparticles for simultaneous determination of hydroquinone and catechol. Analyst, 2013, 138: 3552–3560.
[47] Wang B, Zheng J, He Y, et al. A sandwich-type phenolic biosensor based on tyrosinase embedding into single-wall carbon nanotubes and polyaniline nanocomposites. Sensors and Actuators B-Chemical, 2013, 186: 417–422.
[48] Ghindilis A L, Gavrilova V P, Yaropolov A I. Laccase-based biosensor for determination of polyphenols: Determination of catechols in tea. Biosensorsand Bioelectronics, 1992, 7: 127–131.
[49] Minussi R C, Pastore G M, Duran N. Potential applications of laccase in the food industry. Trends in Food Scienceand Technology, 2002, 13: 205–216.
[50] Fu J, Qiao H, Li D, et al. Laccase biosensor based on electrospun copper/carbon composite nanofibers for catechol detection. Sensors, 2014, 14: 3543–3556.
[51] Amatatongchai M, Sroysee W, Laosing S, et al. Rapid screening method for assessing total phenolic content using simple flow injection system with laccase based-biosensor. International Journal of Electrochemical Science, 2013, 8: 10526–10539.
[52] Chawla S, Rawal R, Kumar D, et al. Amperometric determination of total phenolic content in wine by laccase immobilized onto silver nanoparticles/zinc oxide nanoparticles modified gold electrode. Analytical Biochemistry, 2012, 430: 16–23.
[53] Gamella M, Campuzano S, Reviejo AJ, et al. Electrochemical estimation of the polyphenol index in wines using a laccase biosensor. Journal of Agricultural and Food Chemistry, 2006, 54: 7960–7967.
[54] Stevanato R, Fabris S, Momo F. New enzymatic method for the determination of total phenolic content in tea and wine Journal of Agricultural and Food Chemistry, 2004, 52: 6287–6293.
[55] Ma Y T, Cheung P C K. Spectrophotometric determination of phenolic compounds by enzymatic and chemical methods sbd: A comparison of structure-activity relationship. Journal of Agricultural and Food Chemistry, 2007, 55: 4222–4228.
[56] Lindgren A, Emneus J, Ruzgas T, et al. Amperometric detection of phenols using peroxidasemodified graphite electrodes. Analytica Chimica Acta, 1997, 347: 51–62.
[57] Rosatto S S, Kubota L T, Neto G D. Biosensor for phenol based on the direct electron transfer blocking of peroxidase immobilising on silica-titanium. Analytica Chimica Acta, 1999, 390: 65–72.
[58] Gutes A, Cespedes F, Alegret S, et al. Determination of phenolic compounds by a polyphenol oxidase amperometric biosensor and artificial neural network analysis. Biosensorsand Bioelectronics, 2005, 20: 1668–1673.
[59] Chen H, Li S, Wang S, et al. A new catechol biosensor immobilized polyphenol oxidase by combining electropolymerization and cross-linking process. International Journal of Polymeric Materials and Polymeric Biomaterials, 2013, 62: 620–626.

[60] Makower A, Eremenko A V, Streffer K, et al. Tyrosinase-glucose dehydrogenase substraterecycling biosensor - A highly-sensitive measurement of phenolic-compounds. Journal of Chemical Technology and Biotechnology, 1996, 65: 39–44.
[61] Wollenberger U, Neumann B. Quinoprotein glucose dehydrogenase-modified carbon paste electrode for the detection of phenolic compounds. Electroanalysis, 1997, 9: 366–371.
[62] Burestedt E, Nistor C, Schagerlof U, et al. An enzyme flow immunoassay that uses beta-galactosidase as the label and acellobiose dehydrogenase biosensor as the label detector. Analytical Chemistry, 2000, 72: 4171–4177.
[63] Ghindilis A L, Makower A, Bauer C G, et al. Determination of p-aminophenol and catecholamines at picomolar concentrations based on recycling enzyme amplification. Analytica Chimica Acta, 1995, 304: 25–31.
[64] Szeponik J, Moller B, Pfeiffer D, et al. Ultrasensitive bienzyme sensor for adrenaline. Biosensorsand Bioelectronics, 1997, 12: 947–952.
[65] Mayorga-Martinez C C, Hlavata L, Miserere S, et al. An integrated phenol 'sensoremoval' microfluidic nanostructured platform. Biosensorsand Bioelectronics, 2014, 55: 355–359.
[66] Imlay J A, Chin S M, Linn S. Toxic DNA damage by hydrogen peroxide through the Fenton reaction in vivo and in vitro. Science, 1988, 240: 640–642.
[67] Wardman P, Candeias L P. Fenton centennial symposium-fenton chemistry: An introduction. Radiation Research, 1996, 145: 523–531.
[68] Ishibashi K, Fujishima A, Watanabe T, et al. Generation and deactivation processes of superoxide formed on TiO2 film illuminated by very weak UV light in air or water. Journal of Physical Chemistry B, 2000, 104: 4934–4938.
[69] Kim S, Choi W. Kinetics and mechanisms of photocatalytic degradation of (CH3)nNH4-n+ (0≤ n≤4) in TiO2 suspension: The role of OH radicals. Environmental Scienceand Technology, 2002, 36: 2019–2025.
[70] Liu J, Lagger G, Tacchin P, et al. Generation of OH radicals at palladium oxide nanoparticle modified electrodes, and scavenging by fluorescent probes and antioxidants. Journal of Electroanalytical Chemistry, 2008, 619: 131–136.
[71] Fridovich I. The biology of oxygen radicals. Science, 1978, 201: 875–880.
[72] Hyland K, Auclair C. The formation of superoxide radical anions by a reaction between O2, OH– and dimethyl sulfoxide. Biochemical and Biophysical Research Communications, 1981, 102: 531–537.
[73] Dizdaroglu M, Jaruga P, Birincioglu M, et al. Free radical-induced damage to DNA: Mechanisms and measurement. Free Radical Biology and Medicine, 2002, 32: 1102–1115.
[74] Imlay J A, Linn S. DNA damage and oxygen radical toxicity. Science, 1988, 240: 1302–1309.
[75] Mello L D, Hernandez S, Marrazza G, et al. Investigations of the antioxidant properties of plant extracts using a DNA-electrochemical biosensor. Biosensorsand Bioelectronics, 2006, 21: 1374–1382.
[76] Li P, Zhang W, Zhao J, et al. Electrochemical antioxidant detection technique based on guanine-bonded graphene and magnetic nanoparticles composite materials. Analyst, 2012, 137: 4318–4326.
[77] Fatima Barroso M, Delerue-Matos C, Oliveira M B P P. Electrochemical evaluation of total antioxidant capacity of beverages using a purine-biosensor. Food Chemistry, 2012, 132: 1055–1062.
[78] Barroso M F, Delerue-Matos C, Oliveira M B P P. Electrochemical DNA-sensor for evaluation of total antioxidant capacity of flavours and flavoured waters using superoxide radical damage. Biosensorsand Bioelectronics, 2011, 26: 3748–3754.

[79] Barroso M F, de-los-Santos-Álvarez N, Lobo-Castanon M J, et al. DNA-based biosensor for the electrocatalytic determination of antioxidant capacity in beverages. Biosensorsand Bioelectronics, 2011, 26: 2396–2401.

[80] Barroso M F, de-los-Santos-Álvarez N, Lobo-Castanon M J, et al. Electrocatalytic evaluation of DNA damage by superoxide radical for antioxidant capacity assessment. Journal of Electroanalytical Chemistry, 2011, 659: 43–49.

[81] Qian P, Ai S, Yin H, et al. Evaluation of DNA damage and antioxidant capacity of sericin by a DNA electrochemical biosensor based on dendrimer-encapsulated Au-Pd/chitosan composite. Microchimica Acta, 2010, 168: 347–354.

[82] Xiong H, Wang Y, Zhang X, et al. Evaluation of antioxidative capacity via measurement of the damage of DNA using an electrochemical biosensor and an ionic liquid solvent. Microchimica Acta, 2012, 176: 479–484.

[83] Liu J, Su B, Lagger G, et al. Antioxidant redox sensors based on DNA modified carbon screen-printed. Analytical Chemistry, 2006, 78: 6879–6884.

[84] Liu J F, Roussel C, Lagger G, et al. Antioxidant sensors based on DNA-modified electrodes. Analytical Chemistry, 2005, 77: 7687–7694.

[85] Tian Y, Mao L, Okajima T, et al. Superoxide dismutase-based third-generation biosensor for superoxide anion. Analytical Chemistry, 2002, 74: 2428–2434.

[86] Tian Y, Mao L Q, Okajima T, et al. Electrochemistry and electrocatalytic activities of superoxide dismutases at gold electrodes modified with a self-assembled monolayer. Analytical Chemistry, 2004, 76: 4162–4168.

[87] Beissenhirtz M K, Scheller F W, Viezzoli M S, et al. Engineered superoxide dismutase monomers for superoxide biosensor applications. Analytical Chemistry, 2006, 78: 928–935.

[88] Beissenhirtz M K, Scheller F W, Lisdat F. A superoxide sensor based on a multilayer cytochrome c electrode. Analytical Chemistry, 2004, 76: 4665–4671.

[89] Chen X J, West A C, Cropek D M, et al. Detection of the superoxide radical anion using various alkanethiol monolayers and immobilized cytochrome c. Analytical Chemistry, 2008, 80: 9622–9629.

[90] Tian Y, Mao L Q, Okajima T, et al. A carbon fiber microelectrode-based third-generation biosensor for superoxide anion. Biosensorsand Bioelectronics, 2005, 21: 557–564.

[91] Cortina-Puig M, Munoz-Berbel X, Rouillon R, et al. Development of a cytochrome c-based screen-printed biosensor for the determination of the antioxidant capacity of orange juices. Bioelectrochemistry, 2009, 76: 76–80.

[92] Deng Z, Rui Q, Yin X, et al. In vivo detection of superoxide anion in bean sprout based on ZnO nanodisks with facilitated activity for direct electron transfer of superoxide dismutase Analytical Chemistry, 2008, 80: 5839–5846.

[93] Shpigun L K, Arharova M A, Brainina K Z, et al. Flow injection potentiometric determination of total antioxidant activity of plant extracts. Analytica Chimica Acta, 2006, 573: 419–426.

[94] Ma W, Han D, Zhang N, et al. Bionic radical generation and antioxidant capacity sensing with photocatalytic graphene oxide-titanium dioxide composites under visible light. Analyst, 2013, 138: 2335–2342.

6 Determination of antioxidant capacity by photoelectrochemical method

6.1 Fundamentals of semiconductor optoelectronic chemistry

6.1.1 The energy bands of conductors, semiconductors and insulators

Solid-state materials can be divided into conductors, semiconductors and insulators based on their conductivity. Their mechanism can be described in terms of the filling of electrons in the band.

The solid-state materials can conduct electricity by the directional movement of electrons under the action of the external electric field. As the electric field forces on the acceleration of the electron, both the velocity and energy of the electrons are changed, that is, an exchange of energy occurs between electrons and the electric field. In terms of the energy band theory, the energy change of electrons is their transition from one energy level to another. In the full band, all energy levels are already occupied by the electrons, that is, inner-shell electrons, in an atom. These electrons cannot produce electric current under the action of an external electric field. Hence, they have no contribution to conduction. In the partially filled band, electrons absorb energy from the external field to transition to an unoccupied energy level and form an electric current. And this type of energy band is called conduction band. The bands occupied by the valence electrons in the metal atom are partially filled (Figure 6.1(c)); hence, metals are good conductors. The energy bands of the insulator and the semiconductor are similar. Figure 6.1(a) and (b) shows the full bands that are filled by valence electrons, also known as valence bands. And the middle of these bands is the bandgap and the upper band is the unoccupied band. Therefore, it does not conduct under the external electric field, but this is the case when the absolute temperature is zero. When the external conditions are changed, such as rising temperature or lighting, a less number of electrons in the full band can be excited to the bottom of the unoccupied band, where the electrons will participate in conduction under the action of the field. At the same time, due to the lack of electrons in the full bands, there is some empty quantum state called empty hole near the top of the bands. Thus, the electrons remained in the full bands also play a conductive role. In semiconductors, both the electrons in conduction band and empty hole in valence band are involved in conduction, which is the biggest difference with the metal conductor. The excitation of electrons in insulators needs a considerable amount of energy because of their wide bandgap. At normal temperatures, the electrons that can be excited to the conduction band are few; hence, the conductivity is poor. In semiconductors, many electrons can be excited to the conduction band at normal temperatures because of their

https://doi.org/10.1515/9783110573763-006

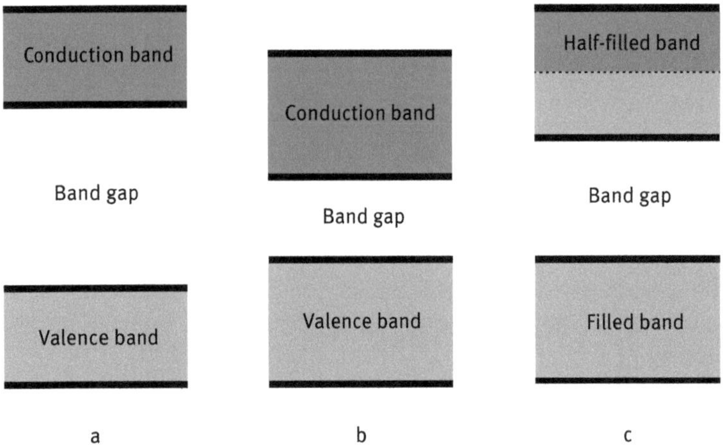

Figure 6.1: Energy band diagram of an insulator (a), a semiconductor (b) and a conductor (c).

small bandwidth (about 1 eV) and certain conductive ability, which is the primary difference between insulators and semiconductors. At room temperature, the bandgap width of diamond is 6–7 eV; hence, it is an insulator, whereas silicon (1.12 eV), germanium (0.67 eV) and gallium arsenide (1.43 eV) are semiconductors. Figure 6.2 shows the energy band of a semiconductor at a certain temperature (in the case of intrinsic excitation), where "●" represents the electron that fills all the energy levels in the valence band at the absolute temperature T = 0 K. E_V is called valence band maximum, which is the highest energy of valence band electrons. At a certain temperature, electrons in the covalent bond can obtain energy which facilitate their escape from covalent bonds to become quasi-free electrons and move freely in crystals by means of thermal or light excitation. The lowest energy required for breaking the covalent bond is the width of band gap (E_g); E_c represents the minimum conduction band, which is the lowest energy of the electrons in the conduction band. The electrons in the covalent bond are excited to quasi-free electrons. In other words, those electrons in the

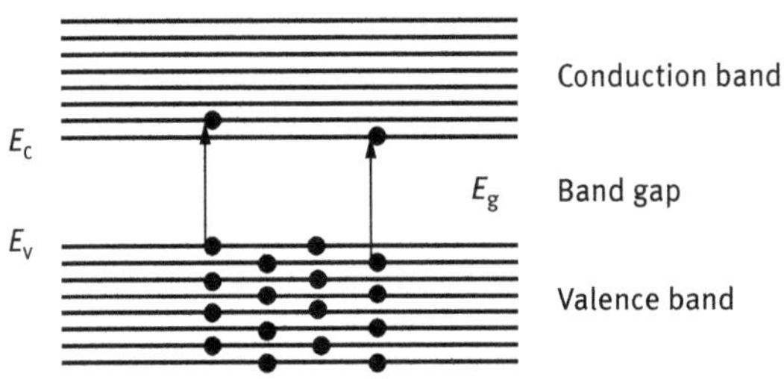

Figure 6.2: Energy band of a semiconductor at a certain temperature.

valence band can be excited up into the conduction band. This process is called intrinsic excitation.

6.1.2 The energy levels of impurities and defects in semiconductors

6.1.2.1 Impurities and defects

In general, various complex phenomena deviate from the ideal situation in the lattice of semiconductors. First, the atom does not rest at lattice sites with a strictly periodic lattice but vibrates around its equilibrium position. Second, the semiconductor material is not pure but contains a number of impurities, that is, there are atoms different from the chemical elements that make up the semiconductor material in the semiconductor lattice. Third, the actual semiconductor lattice structure is not complete, but contains various forms of defects. These defects are generally divided into three categories: (i) point defects such as vacancies and interstitial atoms; (ii) line defects such as dislocations; and (iii) surface defects such as stacking faults and polycrystalline grain boundaries.

Practically, a very small amount of impurities and defects have an influence on the physical and chemical properties of semiconductor materials. Based on the theoretical analysis, it is possible to introduce an energy state (i.e., energy level) into the forbidden band in the presence of impurities and defects or the destruction of the periodic potential field produced by the atoms that are in strict accordance with the periodic arrangement.

6.1.2.2 Substitutional impurity and interstitial impurity

In semiconductors, the impurity atoms exist in two ways. One is that the impurity atoms are located in the interstitial spaces between the lattice atoms, often referred to as interstitial impurities. The other is that the impurity atoms are located at the lattice points instead of lattice atoms, often referred to as substitutional impurities. Interstitial impurity atoms are generally small, for example, the lithium ion (Li^+) has a radius of 0.068 nm; hence, it can be doped in various semiconductor materials. In general, for the formation of substitutional impurities, the size and the valence electron shell structures of the surrogate impurity atoms should be similar to the substituted lattice atoms.

6.1.2.3 Donor impurity and donor level

When a phosphorus atom occupies the position of the silicon atom, the four of five valence electrons of the phosphorus atom form covalent bonds with the surrounding four silicon atoms, leaving one more valence electron. At the same time, the position of the phosphorus atom also has a positive charge $+q$ (the silicon atom has the positive charge $4q$ after removing valence electrons and the phosphorus atom has the positive charge 5*q*), which is called the positive charge phosphorus ion (P^+). Therefore, after the replacement of phosphorus atoms with silicon atoms, there will be a positive center P^+ and extra valence electrons, and the extra valence electrons are tied around the positive center P^+. However, this bound is much weaker than the covalent binding effect, allowing it to break free and produce a free movement for the electrons to conduct in the lattice with a small amount of energy, where the phosphorus atom becomes one less valence electron of phosphorus ions, which are fixed positive center. The above-mentioned process that electrons are separated from impurity atoms to become conductive electrons is called impurity ionization. This kind of impurity that can release electrons during ionization to generate conductive electrons and form a positive charge center is called donor impurity or n-type impurity. The process of releasing electrons is called donor ionization. The donor impurity is neutral when it is not ionized, called a bound state, and becomes a positive charge center after ionization, called the ionization state.

The donor impurity ionization process is shown in a band diagram (Figure 6.3). Electrons move from the donor's bound state to the conduction band and become conductive electrons after getting the energy ΔE_D; hence, the energy of electrons bounded by the donor impurity is ΔE_D which is lower than the bottom of the conduction band E_C. The electron's energy state bound by donor impurity is called donor level, that is, E_D. As ΔE_D is much smaller than E_g, the donor level lies in a forbidden band close to the bottom of the conduction band. Incorporation of donor impurities in pure semiconductors increases the number of conductive electrons in the conduction band after ionization and enhances the conductivity of semiconductors. Semiconductors that rely mainly on the conduction of the conduction band electrons are often referred to as electronic or n-type semiconductors.

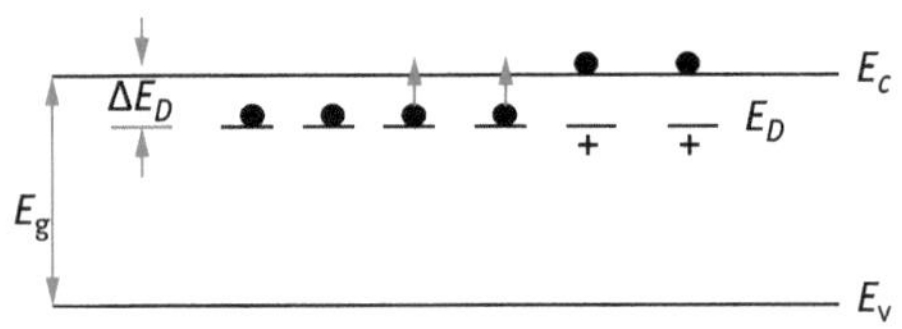

Figure 6.3: Donor level and donor ionization.

6.1.2.4 Acceptor impurity and acceptor level

When the valence electrons of the doped atom are one less than those of the surrounding atoms, it is necessary to take a valence electron from another atom and then a hole is created in the covalent bond of the semiconductor crystal. Impurity atoms that have gained electrons form negatively charged ions as the negative charge center. Negative ions and positively charged holes have an electrostatic attraction; hence, this hole is bound by negative ions and moves around it. However, this binding force is very weak, and holes can get rid of shackles with a small amount of energy moving freely in the crystal's covalent bond. At the same time, this ion is a fixed negative charge center. Impurities that accept electrons from the surrounding crystal atoms to produce conductive holes and form negative charge center are called acceptor impurities or p-type impurities.

The ionization of acceptor impurities is also shown in the band diagram (Figure 6.4). Once the energy ΔE_A is obtained, the hole transitions from the acceptor bound state to the valence band to become the conductive hole. As the energy of the hole in the energy band diagram is higher toward the bottom, the energy of holes bound by the acceptor impurity is lower by ΔE_A than that of the valence band maximum (E_V). The energy state of holes bound by the acceptor impurity is called the acceptor level, that is, E_A. Because E_A is much smaller than E_g, the acceptor level is located in the forbidden zone near the top of the valence band. Of course, the acceptor ionization process is actually the electron movement. Electrons in the valence band transition to the acceptor level, compounding with holes bound to the acceptor's energy level, and a free-moving conductive hole is created in the valence band, forming an immobile acceptor ion. After doping the pure semiconductor, the acceptor impurity is ionized to increase the conduction holes in the valence band and enhance the conductivity of semiconductors. Semiconductors that rely mainly on holes to conduct are often referred to as hole-type or p-type semiconductors.

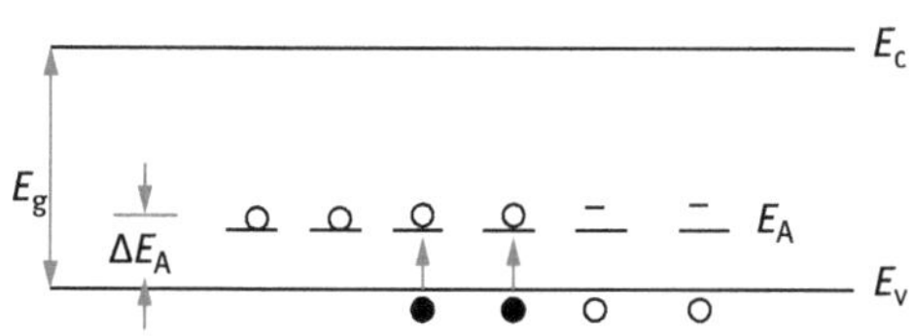

Figure 6.4: Acceptor level and acceptor ionization.

6.1.3 The band edges

The photogenerated electron transport capacity between the semiconductor and the adsorbed species is limited by the location of the semiconductor energy band and the redox potential of the adsorbed species. The relative potentials of the acceptor species are both thermodynamically and kinetically lower (i.e., more positive) than the conduction band of the semiconductor. For the electron donor, on the other hand, its potential level is above the valence band of semiconductors, which is more negative, so that the electron can compensate for the hole in the semiconductor valence band. The relationship between the band edge of the semiconductor and the redox potential of the reactants determines whether photocatalytic reactions can occur, which is the decisive factor.

Figure 6.5 shows the location of the band edge of some semiconductors. The inner energy scale shows that the left edge is under the vacuum condition and the right edge is with respect to the common hydrogen electrode. The measurement of banding is carried out in an aqueous solution of pH 1, and what is measured is the band bending at the level of the bandgap instead of bending point after contact with the measuring electrode. The influence of pH of the electrolyte on the band positions of various semiconductors varies. The band diagram provides the following information:

(1) Semiconductor bandgap is used to estimate the wavelength required to excite semiconductors.
(2) The position of the conduction band determines the reduction ability of the semiconductor.
(3) Valence band position determines the oxidation ability of semiconductors.

These information help us to initially determine whether the redox reaction of semiconductors can occur. In particular, the location of the conduction band has important implications for the determination of antioxidant capacity.

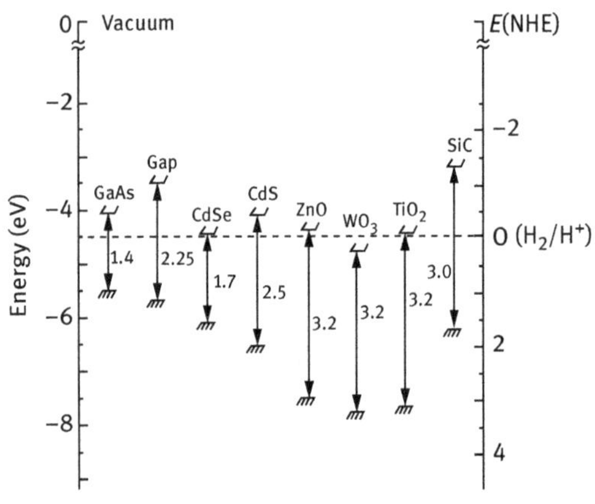

Figure 6.5: Edge position of various semiconductor materials in aqueous solution at pH = 1 [1].

6.2 Functionalization of semiconductor bandgap

If the valence band position of a specific semiconductor is determined, its theoretical redox ability is also determined. However, in the actual application process, often the existence of the natural conduction band and valence band positions cannot meet the demand, especially in the realization of photoelectrochemical method analysis and detection; hence, the band structure of a semiconductor material needs to be adjusted. In recent years, extensive and in-depth studies have been conducted on the modification of semiconductors, such as doping, semiconductor recombination, noble metal loading and surface plasmon photocatalysis to broaden the light absorption of the semiconductor and increase the intensity of light absorption. In combination with the photocatalytic principle, adjusting the semiconductor energy band by changing the semiconductor electronic structure is the most effective means to expand the visible light response of a material. This chapter summarizes the research progress of semiconductor modification in recent years based on the three aspects: doping of semiconductors, formation of solid solution and semiconductor composite.

6.2.1 Semiconductor doping

6.2.1.1 Metal element doping

One of the most effective means of preparing a visible light photocatalytic material is doping of metal elements to introduce an impurity level in the middle of the forbidden band. This method has been used to develop the light absorption of a semiconductor photocatalytic material with wide bandgap. Over the past few decades, there have been many studies on the metal element doping in semiconductors, especially semiconductors with wide bandgap. In particular, there is progress in metal element doping on oxide semiconductors such as titanium dioxide (TiO_2).

Figure 6.6 shows the effect of metal doping on the energy band of a semiconductor. When the metal element is doped, it can generate a donor level above the valence band or an acceptor level below the valence band. When the semiconductor is exposed to light, in addition to the electron excitation from the valence band to the conduction band, the electron undergoes transition from the donor level to the conduction band

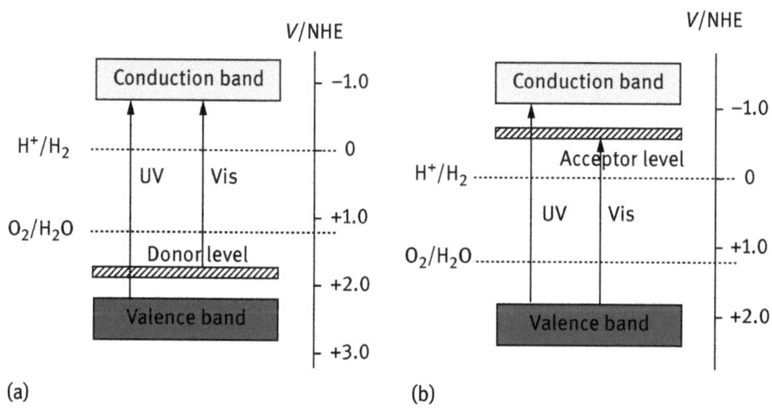

Figure 6.6: Effect of metal doping on the energy band of the semiconductor [2].

or from the valence band to the acceptor level resulting from the doping. Then the absorption range of the semiconductor and the potential of electron–hole pairs generated after the excitation are changed.

Thus far, TiO_2 is the most widely studied material in the field of doping metal ions. TiO_2 doped with V, Ni, Cr, Mo, Fe, Sn and Mn has been reported. Umebayashi et al. [3] systematically studied electronic structures of TiO_2 doped with various metals (Figure 6.7) by first-order theory calculations. They found that after doping transition metals, the t2g orbital occupancy can be generated in the middle of the forbidden band, and electrons can transition between the t2g level and the valence band (conduction band). Some studies assumed that the doping of V, Cr, Mn, Fe, Co, Ni and Rh

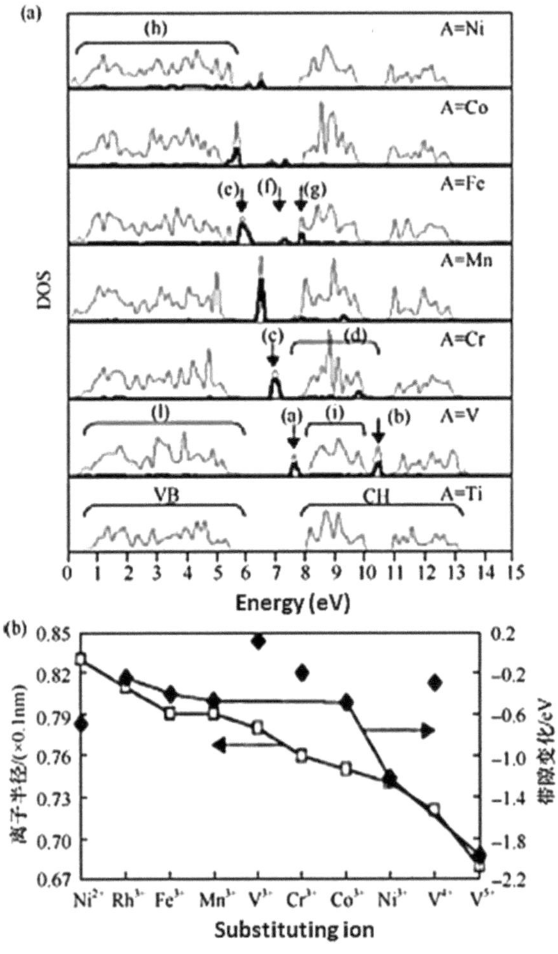

Figure 6.7: (a) The distribution of electronic density of states of TiO_2 doped with different metals and (b) the relationship between the radius of metal ions and the change of energy band [3].

all can cause the red shift of the absorption edge of TiO_2, and the ion radius also has an influence on the band regulation. The doping of Ni and V ions is the most effective to shorten the energy gap [4]. Some studies also state that the lanthanide doping can significantly improve the photocatalytic activity of TiO_2. Since the electrons in f electron orbitals of lanthanides is apt to complex with the Lewis acid, the doping of lanthanides into semiconductors will greatly facilitate the enrichment of Lewis acid base organic compounds at TiO_2 surface. TiO_2 doped with La^{3+}, Eu^{3+}, Nd^{3+} and Ce^{4+} lanthanides show some visible photocatalytic activity. V doping is considered as one of the most effective ways to expand the spectral response range of TiO_2 because V can effectively improve the separation efficiency of photogenerated carriers. In V-doped TiO_2, V^{4+} can serve as the electron–hole capture center, while V^{5+} is the electron acceptor, effectively increasing the photocatalytic quantum yield [5].

In general, the incorporation of appropriate amount of metal ions in TiO_2 can effectively prevent the recombination of the photoelectron–hole pair. Meanwhile, substitution of Ti^{4+} by metal ions can reduce the energy band of TiO_2 and make it absorb visible light. A large number of studies have reported that metal doping can indeed cause the absorption of visible light, including noble metal doping, rare earth ion doping and transition metal doping. Especially precious metals such as gold and silver, due to their own surface plasmon effect, can absorb visible light; hence, the doping of these precious metals can further enhance the visible light activity of TiO_2. However, some studies suggest that the doping of metal ions does not act as the capture center of photogenerated carriers, but increases only the number of recombination centers of photogenerated carriers, leading to the decrease in catalytic efficiency.

In summary, the metal element doping can effectively change the semiconductor electronic structure, thereby increasing the visible light absorption range of semiconductors. The types of doping elements, the doping method, the doping amount and the doping position can all cause the change of the photocatalytic activity of the semiconductor. Some controversy still exists over whether the doping of metal elements can improve the photocatalytic activity of semiconductors.

6.2.1.2 Nonmetallic element doping

Nonmetallic doping is another way to adjust the energy band of the ultraviolet (UV)-responsive photocatalytic material. Nonmetallic elements such as (N, S, V, C)-doped TiO_2 have been widely studied. Unlike the doping of metallic elements, the doping of nonmetallic elements does not form the donor level in the middle of the forbidden band, but the band can be reduced by hybridization of electrons in the same valence band (Figure 6.8). Nonmetallic doping has been widely used to control the UV-responsive semiconductors with wide bandgap such as Ti-based oxide, Ta-based oxide and Zr-based oxide.

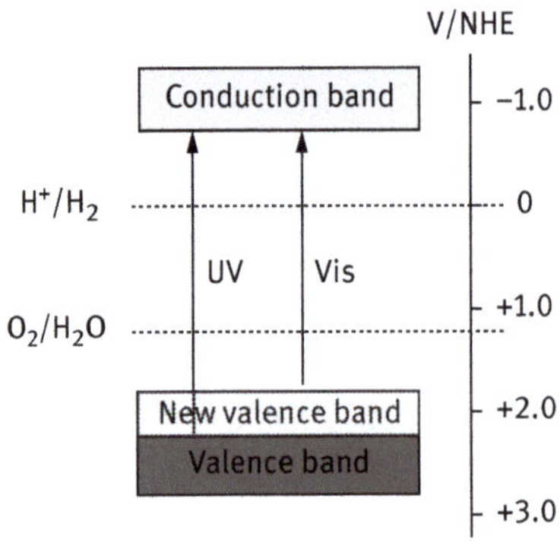

Figure 6.8: Effect of nonmetallic element doping on the energy band of the semiconductor [2].

A wide variety of nonmetallic ions are used to dope TiO_2 to tune its energy band. Relative to undoped TiO_2, the absorption band side after doping nonmetallic element will red shift to long wavelength. Chen et al. conducted a detailed study of C-, N- and S-doped TiO_2 by X-ray photoelectron spectroscopy (XPS). As shown in Figure 6.9(a), these doped ions give rise to new peaks in the photoelectron spectra, and these extraneous states lead to the red shift of the absorption band edge (Figure 6.9(b)) [6, 7]. Asahi studied the effect of oxygen atoms on the band structure of TiO_2 substituted by N, C, F, P and S. Full-potential linearized augmented plane wave method was used to calculate the energy band after the substitution of TiO_2 by different elements. It was found that the 2p orbitals of N easily intermixed with the 2p orbitals of O, resulting in the increase in valence band, shortening the forbidden bandwidth of TiO_2 and increasing the absorption of visible light [8].

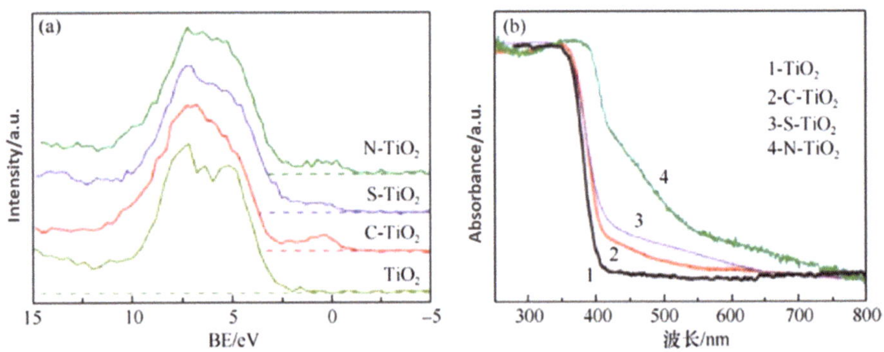

Figure 6.9: (a) XPS spectra of titania doped with different elements and (b) the corresponding UV-Vis diffuse reflectance spectra [7].

Yet, there are many ways to perform N doping of TiO_2, including vapor deposition (chemical and physical), simultaneous heating of TiO_2 and nitrogen-containing organic compounds (urea), magnetron sputtering, hydrothermal / solvothermal methods and anneal of TiO_2 under NH_3 atmosphere. Similarly, the study on other nonmetallic doped TiO_2 is also quite extensive. Through different synthesis methods, element S can either replace the position of O in the form of anion, or Ti in the form of cation [9, 10]. Moreover, a variety of chemical synthesis methods can achieve the C element doping of TiO_2, and can form multiphase mixed (anatase, rutile and brookite) TiO_2 samples. In order to further improve the catalytic activity of visible light, TiO_2 and other semiconductors were double-nonmetal doped, such as F/B codoping, F/N codoping, F/S codoping, F/C codoping, C/N codoping, S/N codoping, N/Br codoping, N/B codoping, N/Si codoping, C/S codoping and F/P codoping [11].

6.2.1.3 Codoping of metal and nonmetallic elements

Codoping of metals and nonmetals is a combination of the earlier two methods to simultaneously raise the valence band position and introduce impurity levels to change the bandgap and band structure of the material. The principle is similar to the doping of metallic and nonmetallic elements. In photocatalysis, the codoped method is mostly used in the photocatalytic oxidation of pollutants, seldom in the photocatalytic cracking of water to produce hydrogen. Codoping is indeed a very effective method in adjusting the energy band, and in some cases it is suitable to use this method in photoelectric method detection and analysis. Gai et al. doped TiO_2 by charge compensation of donor–acceptor pairs such as (N + V), (Nb + N) and (Cr + C) [12, 13]. The photocatalytic decomposition of water by (Mo + C)-doped TiO_2 is the most effective, but the improvement of conversion efficiency did not come from the change in the conduction band position. Lei et al. used hydrothermal methods to synthesize sulfur-substituted and Zn-doped $In(OH)_3$ [12, 13]. As shown in Figure 6.10, the energy gap of In $(OH)_3$ becomes narrower because S^{2-} replaces OH^-. The valence band of this material is mainly formed by the hybridization of 3p orbit of S and the 2p orbit of O. At the same time, the valence band composed of 5s5p orbitals of In and 4s4p orbitals of Zn increases the valence band position due to the introduction of Zn^{2+}, which provides a greater thermodynamic driving force for the reduction of H_2O to hydrogen.

6.2.2 Formation of a solid solution

In addition to the doping of heterogeneous elements, the formation of a solid solution with a wide bandgap semiconductor and a narrow bandgap semiconductor is also an effective way to change the electronic structure and the band structure of semiconductors. In solid solution, ions are replaced at a certain crystalline structure of a mineral without changing the structure and symmetry of the entire crystal, that is, the solute atoms dissolve into the solvent crystal lattice while still maintaining the solvent type of alloy phase. Hence, the solid solution is a kind of isoelectronic doping of semiconductors with high concentration. Once the solid solution is formed from the two semiconductors, the change in the composition of an element causes the change in the corresponding electronic structure without changing the structure. As shown in Figure 6.11, by changing the compositional ratio of the solid solution, the energy band position and the bandgap of the material are changed simultaneously.

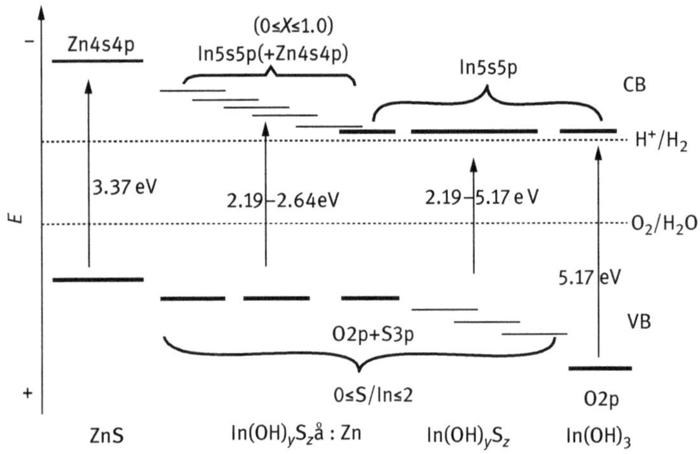

Figure 6.10: Theoretical valence band and conduction band structures of $In(OH)_3$, $In(OH)_yS_z$ and $In(OH)_yS_zZn$ [12, 13].

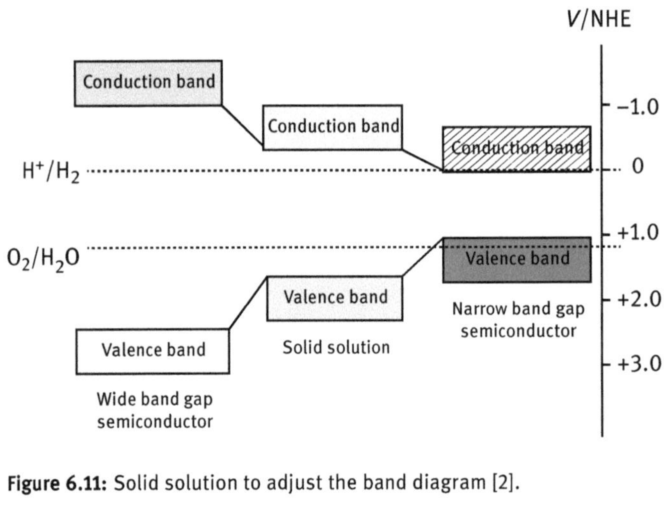

Figure 6.11: Solid solution to adjust the band diagram [2].

6.2.2.1 Solid solution of sulfur compound

By combining wide bandgap ZnS and narrow bandgap CdS in different proportions, $Cd_xZn_{1-x}S$ solid solution is obtained in the visible light photocatalytic field, which led to a considerable amount of research and application. The solid solution $Cd_xZn_{1-x}S$ prepared by coprecipitation has a good activity of photolysis and hydrogenation of hydrogen under the visible light in the system by adding S_2^- and SO_3^{2+} as sacrificial agents. Zhang et al. [14] prepared the $Cd_xZn_{1-x}S$ solid solution by thermal Zn and Cd oxide precursors and found that the photocatalytic activity of the obtained sample was higher than that obtained by the coprecipitation method. Especially when the value of x is about 0.2, the quantum efficiency of hydrogen production reaches 10.23% at 420 nm.

Since 2004, Kodo's group has done a considerable number of research works on solid solutions of ZnS and $MInS_2$ (M = Cu, Ag). The band structure of the solid solution can be adjusted by controlling the ratio of ZnS and $MInS_2$ [15]. For example, in ZnS–$AgInS_2$, as shown in Figure 6.12, the diffuse reflectance UV–visible absorption band edges and the positions of conduction band and valence band differ with the change in composition. It can be seen from the figure that the position of the energy band can be controllably adjusted by making the proportion of each substance in the solid solution to obtain the excitation energy needed for the photoelectric detection and analysis.

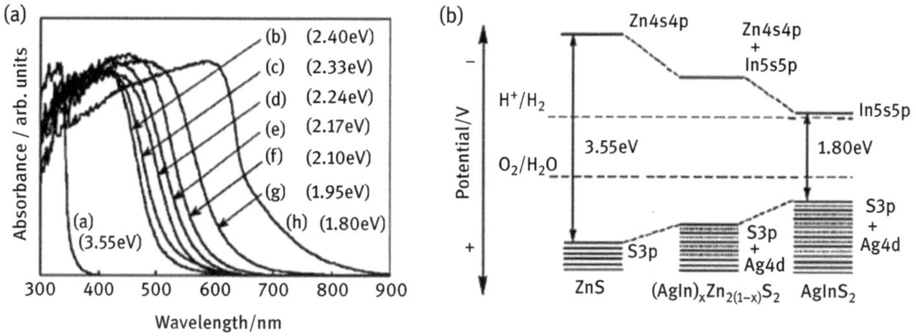

Figure 6.12: (a) $(AgIn)_xZn_{2(1-x)}$ S_2 solid solution diffuse reflectance absorption changes of x values were (1) 0, (2) 0.17, (3) 0.22, (4) 0.29, (5) 0.33, (6) 0.44, (7) 0.5, (8) 1; and (b) change in energy band diagram. [15].

6.2.2.2 Oxide solid solution

Studies on solid solutions formed from oxide semiconductors with similar crystal structures are equally extensive. Kudo and Mikami studied the solid solution formed from In_2O_3 and Ga_2O_3 [16], including light absorption, luminescent properties and band position. In addition, they also prepared a solid solution of $Sr_2(Ta_{1-x}Nb_x)_2O_7$ and found that the energy bands of solid solution with this layered perovskite structure changes with the ratio of Ta and Nb, but all the width of samples is greater than 3.9 eV. Moreover, Zou et al. also prepared solid solutions of $BiTa_{1-x}Nb_xO_4$ ($0 < x < 1$) by solid-state reaction [17]. Although this oxide solid solution can only have photoactivity under UV light, it also provides us a method to adjust the energy band of the oxide semiconductor.

Luan et al. obtained a solid solution of $Bi_xIn_{1-x}TaO_4$ ($0 < x < 1$) by solid-phase reaction [18]. When x is equal to 0.2, 0.5 and 0.8, the energy bands of the solid solution are 2.86, 2.71 and 2.74 eV, respectively. The conduction band of the solid solution is mainly composed of 5d orbitals of Ta, 5p orbitals of In and 5s orbitals of In, whereas the valence band is composed of 2p orbitals of O, 6s orbitals of Bi and 6p orbitals of Bi. The solid solution becomes excited and exhibits a certain degree of photoactivity under visible light irradiation with a wavelength greater than 420 nm. Visible light absorption of the sintered $Na_{1-x}La_xTa_{1-x}Co_xO_3$ solid solution is obtained mainly by the hybridization of 3d orbitals of Co with 2p orbitals of O [19]. There are many ways and combinations (sulfide, oxide, nitride and halide) to form a solid solution, but only a few of the types are listed here. In summary, the use of solid solution methods to adjust the position and structure of the band is a better way to control the band position by adjusting the ratio, which is theoretically easier. The basic principle in choosing a solid solution combination is to consider both the absorbance range (bandgap) and the redox ability (band position) of the sample.

6.2.3 Semiconductor composite

6.2.3.1 Sensitization

The system is a combination of a narrow bandgap semiconductor and a wide bandgap semiconductor. The photogenerated electrons are excited to the conduction band by visible light in a narrow bandgap semiconductor. Due to their slightly higher conduction band than the wide bandgap semiconductors, photogenerated electrons spontaneously transfer to the conduction band of the wide bandgap semiconductor, while the holes remain in the valence band of the narrow bandgap semiconductor. As shown in Figure 6.13, such a structure facilitates the separation of electron holes and reduces their recombination. In the case of visible light excitation, the catalytic conduction band of this system is the conduction band of a wide bandgap semiconductor, and the catalytic valence band is the valence band of a narrow bandgap semiconductor. The most common type of composite system is the sensitization of chalcogenide semiconductor materials to oxide wide bandgap semiconductor materials. As oxide semiconductors have good catalytic activity and the wide bandgap cannot be easily excited, the chalcogenide semiconductor is recombined with it, which can achieve visible light absorption under the condition of ensuring its catalytic ability, such as $CdS–TiO_2$ system, [20] $Cd_{0.5}Zn_{0.5}S$ system [21] and $CdSe–TiO_2$ system [22].

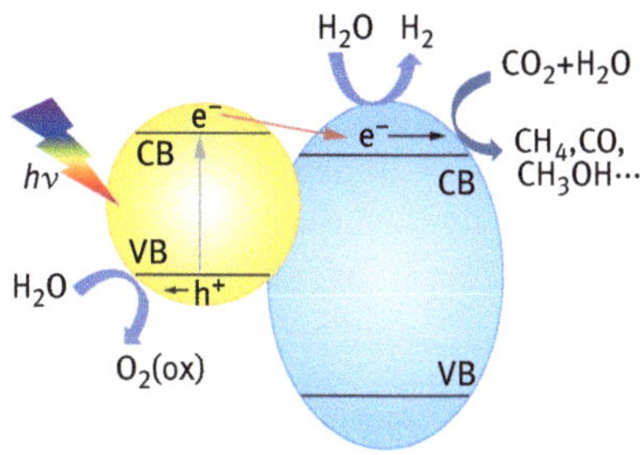

Figure 6.13: A schematic of electron transfer sensitization mechanism in semiconductor composites [23].

6.2.3.2 p–n heterogeneous section

The structure of the system is similar to the one of diode, which consists of an n-type semiconductor and a p-type semiconductor. Both of them can be simultaneously excited to generate their own electron–hole pairs [24]. After excitation, the electrons transfer from the higher conduction band to the lower conduction band, while holes transfer from the lower conduction band to the upper conduction band, which is shown in Figure 6.14. This is conducive to the carrier separation, which will facilitate the catalytic reaction. In this system, two semiconductors whose energy gaps are not very wide are selected and are excited by light. The electrons and holes are enriched in two materials, respectively. Photoelectric efficiency is higher, but the material's overall energy gap will be narrower than the two materials, namely, the conduction bands give first place to the lowest while the valence bands give first place to the highest. There are not many studies on this compound method. Some of the examples are: p-type semiconductor $CaFe_2O_4$, n-type semiconductor $MgFe_2O_4$ with bandgaps of 1.9 and 2.0 eV, respectively [25], ($Cr\text{-}Ba_2In_2O_3$)–($Cr\text{-}In_2O_3$) composite system [26] and CdS and TaON core–shell structure composites [27].

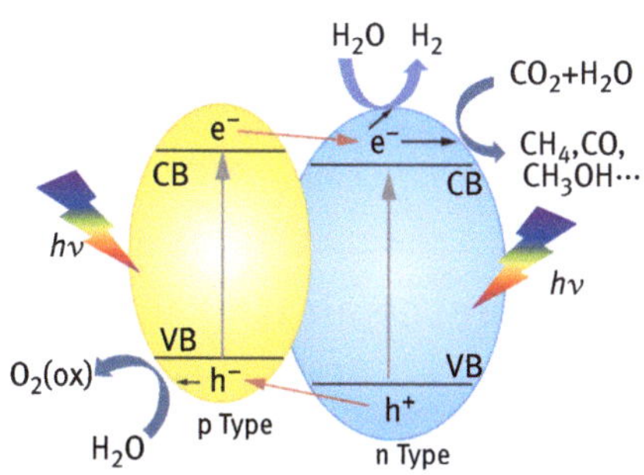

Figure 6.14: Schematic diagram of electron transfer for p–n heterostructure mechanism in semiconductor composites [23].

6.2.2.3 Z-type combination

For a semiconductor catalyst, the edge of the valence band and the edge of the conduction band should match the redox potential of the reactant. For example, the conduction band in the photolysis water reaction should be higher (more negative) than the redox potential of H_2/H_2O, while the valence band should be lower than the O_2/H_2O potential. However, some semiconductors cannot satisfy both of these potentials at the same time. In such cases, a Z-type combination of the two semiconductors is required. After being excited at the same time, the photogenerated electrons of the semiconductor with the lower conduction band will recombine with the photogenerated holes of the higher valence band semiconductor, leaving more negative photogenerated electrons and more positive photogenerated holes. The process of electron transfer is shown in Figure 6.15. The advantage of this structure is that it is possible to utilize both the reduction of the higher conduction band and the oxidation of the lower valence band. Nowadays, this method is the most popular one to adjust the band structure.

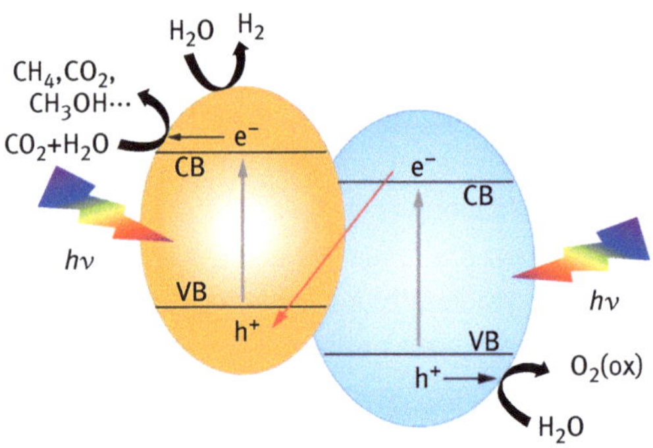

Figure 6.15: A schematic of electron transfer mechanism of a semiconductor composite Z-type combination [23].

We can use one or several composite methods to obtain the required positions of conduction band and valence band, and then for the photoelectric detection and analysis applications.

6.3 Basic principles of semiconductor photochemical detection

The properties of semiconductor electrodes and metal electrodes are very different, based on their nature. In electrochemical measurement, this difference is mainly manifested in the potential drop. When a certain bias voltage, especially reverse biased, is applied to the semiconductor electrode, or when the intrinsic Fermi level of the semiconductor is in the vicinity of the middle of the forbidden band, potential changes at the semiconductor/electrolyte interface cannot occur at Helmholtz layer, but inside the semiconductor near the surface.

According to the band theory, the Fermi level (E_f) of semiconductors lies between the bottom of the conduction band (E_c) and the top of the valence band (E_v). When the semiconductor is in contact with the solution, charge transfer occurs at the interface due to the difference between the semiconductor E_f and the solution potential (φ_{soln}). For example, in n-type semiconductor, if its E_f is higher than the solution potential, most of the carriers (electrons) in the semiconductor are transferred to the solution, so that the inside of the semiconductor is positively charged and band bending occurs. The potential barrier formed by the upward band bending prevents the electrons in the semiconductor from transferring to the solution until equilibrium is reached, and the potentials in the two phases are equal (the interface between the formed semiconductor and the solution is shown in Figure 6.16). The concentration of carriers in semiconductors is very low, that is, about 10^{13}–10^{16} cm^{-3}. Once the semiconductor is in contact with the solution, a residual charge layer of 0.1–100 μm, that is, a space charge layer, is generated on the side of the semiconductor. When the interface (space charge layer) between the semiconductor and the solution is exposed to photons of energy greater than the bandgap of the semiconductor, the electrons of valence band are partially excited to the conduction band. Under the electric field of the space charge layer, most charge carriers move to the inside of the semiconductor, and minority charge carriers move to the semiconductor surface; hence, the separation of photogenerated carriers occurs. The photogenic hole can participate in the interface reaction.

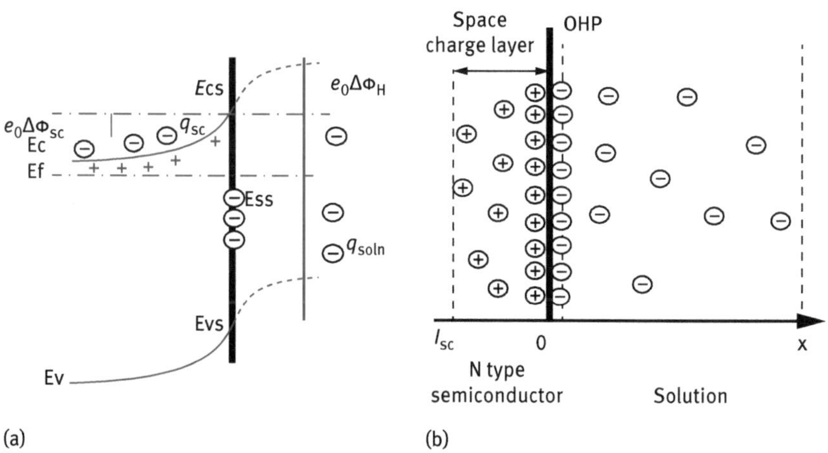

Figure 6.16: Energy band bending (a) and space charge layer (b) at the semiconductor–solution interface.

In the actual semiconductor photoelectrocatalytic system, even though the photogenerated carriers are separated by the space charge layer, their kinetics are still very complex. The bold arrows in Figure 6.17 [28] represent the process of reducing the substance in the hole oxidation solution and introducing electrons into the counter electrode, that is, an effective photocurrent. All other processes are recombination or reverse current (inactive photocurrent) that occurs during the photoexcitation in the final reaction, which can be regarded as resistance in the process of charge transfer, resulting in the reduction of photocurrent. These include J_{br}, J_{dr}, J_{ss}, J_t and J_{et}. The various processes can be refined through further analysis. Bulk-phase recombination and depletion-layer recombination are not only related to semiconductor crystallinity, bulk defects and carrier mobility, but also include charge transfer between semiconductor particles and the substrate. The complex surface states include the surface state of the semiconductor, the surface state formed by contact with the solution, the photoelectrocatalytic intermediate state and the surface state caused by the product adsorption. Electron tunneling current and electrons over the barrier current belong to the diffusion current of electrons. To increase the value of photocurrent, various methods must be used to avoid the generation of noneffective photocurrent.

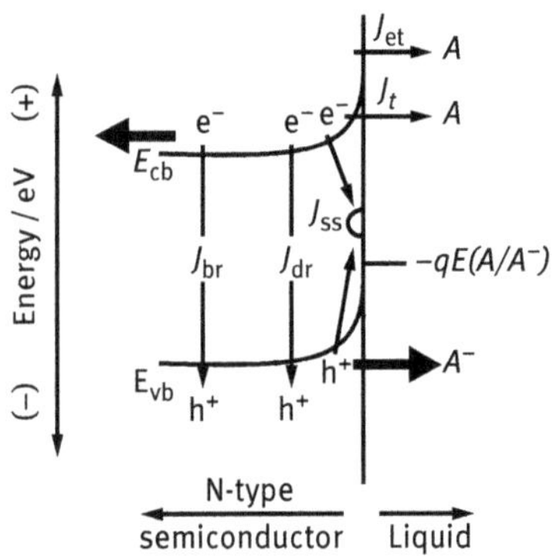

Figure 6.17: Photogenerated carrier-based dynamics at n-type semiconductor–solution interface [28].

In the actual photoelectrocatalytic system, Uosaki et al. found that the photocatalytic reaction rate (photocurrent) and the electrode potential are in accordance with the Tafel relationship when the photocurrent is lower than the saturation current, indicating that the kinetic velocity step at this time is at the interface kinetic process, rather than the photogenerated carrier transport within the semiconductor [29]. In other words, the surface state of the semiconductor has a major impact on photocurrent. Then high-density surface-state photocurrent theory, that is, Helmholtz barrier model is introduced [30].

The experimental curves of photocurrent and applied bias voltage and the theoretical curves obtained according to the Helmholtz barrier model are shown in Figure 6.18. Both the curves are basically the same. The theory of photocurrent under monochromatic light (v) is expressed as

$$J_{ph}(v) = \frac{Ae_0 I_0 k_{ct}}{k_{ct} + k_{sr} + k_{br}}$$

where I_0 is the light intensity absorbed by the semiconductor, A is the space charge layer-related parameter, k_{ct} is the electron transfer reaction rate constant on the electrode surface and k_{sr} is the recombination rate constant of the electrode surface. The sum of the electron transfer reaction and the surface recombination is the total amount of electrons delivered to the electrode surface. As shown in Figure 6.18, 0.4 V indicates the turn-on potential of the photoelectrode. Since k_{sr} and k_{br} approach infinity at this time, the photocurrent approaches zero. In the 0.4–0.4 V region, $(k_{sr} + k_{br}) \gg k_{ct}$, the charge transfer becomes the rate-determining step. The photocurrent at this time is determined by three factors. When the saturation current is less than −0.4 V, $k_{ct} \gg (k_{sr} + k_{br})$, the reaction is controlled by the carrier transport of the semiconductor.

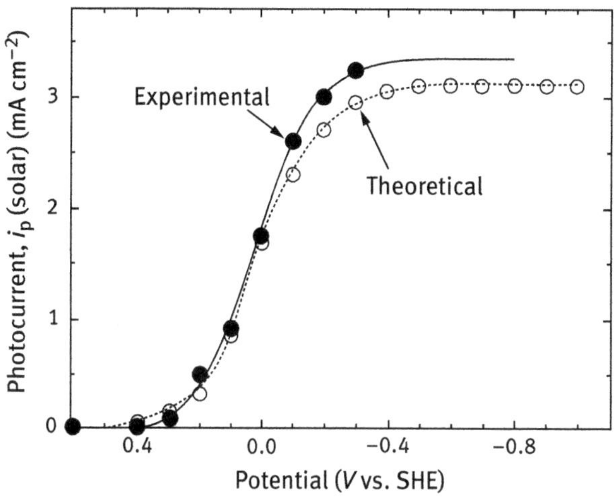

Figure 6.18: Theoretical simulation of photocurrent curve [30].

By the photoelectrochemical method, a type of photoelectrocatalytic method, the antioxidant capacity of antioxidants is measured. This process is shown in Figure 6.19. Under suitable light irradiation, the n-type semiconductor electrons transition from the valence band to the conduction band and produce electrons (e^-) and holes (h^+). Owing to the action of the space charge layer, the electrons flow to the internal electrode to generate photocurrent, whereas holes move to the surface electrode to participate in the oxidation reaction. Then the holes react with the added antioxidant. The electrons and holes are further excited by the light, resulting in a stronger photocurrent. The antioxidant capacity is measured based on this principle.

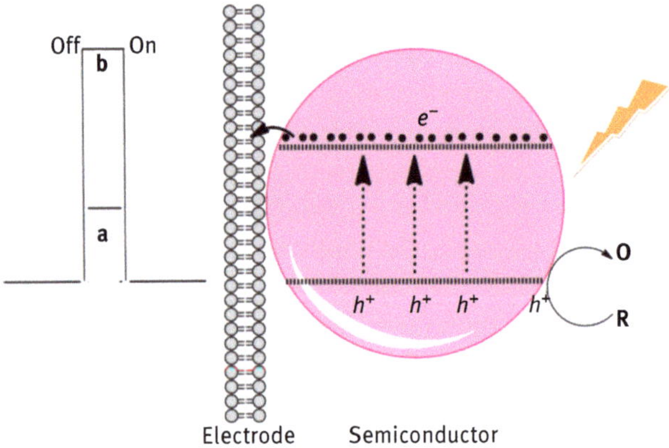

Figure 6.19: Mechanism of antioxidant capacity by photoelectrochemistry. (a) Without antioxidant and (b) with antioxidant.

Three key points in the analysis and applications of photoelectric principle, especially in the antioxidant capacity detection and analysis of antioxidants, are: (1) Sensitivity and detection limit. After selecting the semiconductor, it is necessary to ensure the effective current value as large as possible to attain high sensitivity and detection limit; that is, as much as possible, holes and reaction electrons of antioxidants are shown as the effective photocurrent. (2) Stability: To ensure the stability of the measurement, which will express as invariable photocurrent, the semiconductor should be resistant to photocorrosion. And the resulting product during determination should be promptly removed to avoid the formation of the surface state or decrease the effective electrode area by resultant covering. (3) Anti-interference: To a large extent, the valence band determines the semiconductor's oxidation capacity; hence, the appropriate valence band position is particularly important. It is necessary to react with all antioxidants, but can not react with carbohydrates and amino acids in food. To meet the above requirements, a natural semiconductor often fails to accomplish the required task, requiring two or more kinds of recombination and functionalization of the semiconductor bandgap.

6.4 Development of several photoelectrochemical antioxidant capacity sensors

6.4.1 Antioxidant capacity determination of a photoelectrochemical antioxidant

The total antioxidants are measured to reflect the synergistic effect among the antioxidants in the food and optimally demonstrate their overall antioxidant capacity. However, with the development of antioxidant research, determination of the antioxidant capacity of an antioxidant should not be overlooked, because it is the basis for determining the antioxidant fingerprint. Antioxidant fingerprints reflect the properties of each component in foodstuffs; hence it is very useful for antioxidative fingerprinting analysis.

The researchers used photochemical methods to determine gallic acid (GA) [31]. GA is an important natural antioxidant, which is widely found in plants such as tea, grape, gallnut, witch hazel and lacquer tree. The study found that GA can effectively eliminate the hydroxyl radicals with high activity in vivo to show antihistamine, anti-inflammatory and antitumor activities and can effectively inhibit the onset of various cancers. Therefore, research in the rapid and sensitive determination of the antioxidant capacity of GA is of great significance. At present, several common methods have been applied to the detection of GA, such as cyclic voltammetry, chemiluminescence and high-performance liquid chromatography-mass spectrometry, but most of these methods require complex sample preparation, a longer time, energy needs and higher costs. Therefore, the search for a portable, inexpensive and sensitive method for determining the antioxidant capacity of GA still faces a huge challenge.

The key of a photoelectrochemical sensor is the choice of photoelectric materials. An excellent photoelectric material should have the following advantages: suitable visible light absorption that can ensure higher photocurrent in visible light to obtain ideal sensitivity; nontoxicity and easy to obtain; and the appropriate valence band position that allows it to react with GA but not with disruptors of sugars and amino acids. Due to its good photostability, biological and chemical inertness, nontoxicity and high carrier concentration, TiO_2 has become the most widely used photocatalytic material at present, but a wider bandgap (3.2 eV) makes it possible to absorb only UV light, which cannot utilize the visible light that occupies most of the sunlight, limiting its photocatalytic efficiency. In recent years, conductive polymer polythiophenes, polypyrrole, poly(*p*-phenylene vinylene) and their derivatives have been widely used to modify TiO_2 in order to overcome this disadvantage. Conductive polymer polyaniline (PANI) is used to modify TiO_2 because of its good conductivity, strong visible light absorption, high absorption coefficient and so on, thereby greatly improving the photoelectric properties of TiO_2. Even more gratifying is that the conductive polymer PANI also has advantages of light corrosion resistance, low toxicity, suitable valence band position and higher carrier concentration, which itself shows good optoelectronic properties. In addition, graphene has a large number of π–π

conjugated structure; hence, as a good electronic conductor, it can reduce the electron and hole recombination and is widely used in the field of photocatalysis. Researchers combined the advantages of TiO_2, PANI and graphene to synthesize reduced graphene oxide (rGO) /TiO_2/PANI composites. By using this photoelectrocatalytic material, a photoelectrochemical sensor was successfully constructed and applied to evaluate the antioxidant capacity of GA. The optoelectronic platform shows the advantages of rapid response, high sensitivity and high selectivity.

6.4.1.1 Structure and morphology features of materials

Using $TiCl_3$ and GO (graphene oxide) as starting materials, rGO–TiO_2 was first synthesized by a simple method and then compounded with different proportions of PANI (100:0.75, 100:1.5, 100:3, 100:12). Finally, rGO/TiO_2/PANI was obtained. Synthetic materials were subsequently characterized using different means. As shown in Figure 6.20(a), TiO_2 nanoparticles of about 10 nm are uniformly distributed on the reduced graphene sheets, which facilitate the electron transfer. The X-ray diffraction (XRD) results (Figure 6.20(d)) show that the 2θ angles exhibit characteristic peaks at

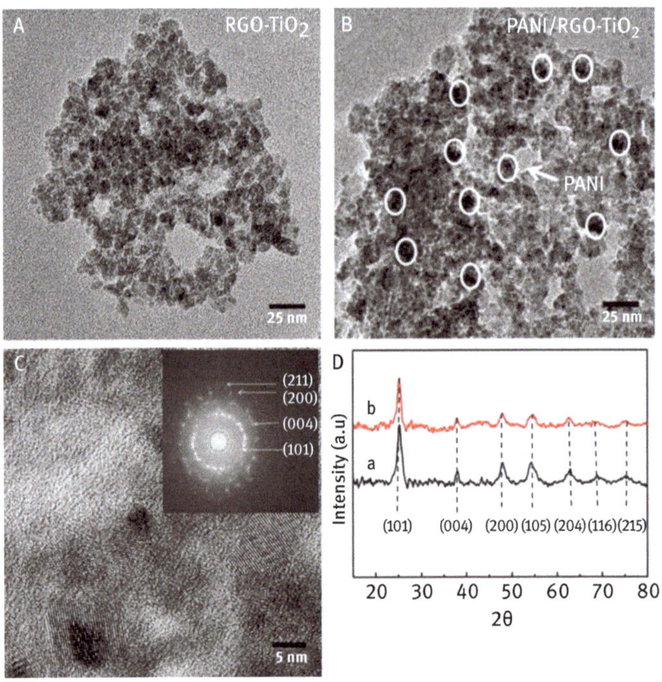

Figure 6.20: Transmission electron microscopic (TEM) images of (A) rGO–TiO_2 and (B) rGO/TiO_2/PANI; (C) high-resolution TEM (HRTEM) image of GO/TiO_2/PANI; (D) XRD pattern of rGO–TiO_2 (a) and rGO/TiO_2/PANI (b).

25.3°, 37.8°, 48.0°, 53.9°, 55.1°, 62.7°, 68.8°, 70.3° and 75.0° representing anatase-(101), (004), (200), (105), (211), (204), (116), (220) and (215) crystal planes. This illustrates that anatase with high catalytic activity was synthesized, which was further confirmed by the high-resolution image of rGO–TiO_2 (Figure 6.20 (c)). When PANI was introduced, no change in the 2θ angle of TiO_2 was found, indicating that PANI did not affect the crystalline form of TiO_2. The presence of C, O, N and Ti elements is clearly visible on the EDS (energy-dispersive spectrometer) pattern of rGO/TiO_2/PANI (Figure 6.21 (b)), which further demonstrates the successful synthesis of rGO/TiO_2/PANI material.

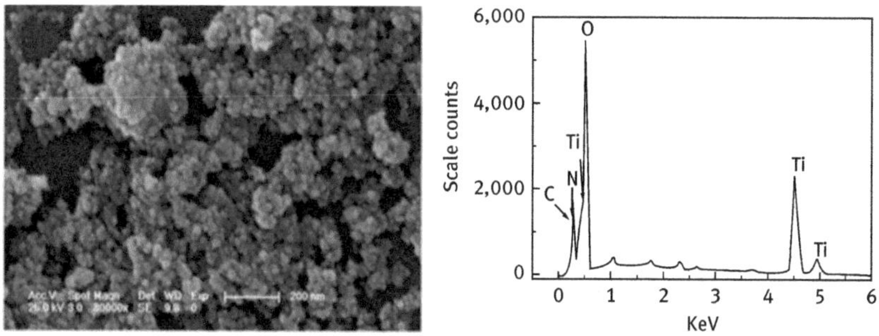

(a) (b)

Figure 6.21: SEM image of rGO/TiO_2/PANI (a) and EDS of rGO/TiO_2/PANI (b).

6.4.1.2 rGO/TiO_2/PANI optical signal response

The photoelectric sensors were designed and constructed with TiO_2, rGO–TiO_2 and rGO/TiO_2/PANI as optoelectronic materials, respectively. As shown in Figure 6.22(a), under visible light irradiation, the blank photocurrent responses were 13.6 and 75.1 nA on the sensor with TiO_2 and rGO–TiO_2 as the optoelectronic material. The introduction of rGO increased the photocurrent nearly fivefold, which was mainly attributed to two factors. First is the introduction of rGO, which increased the light absorption in the UV–visible region of the composite (Figure 6.23 (a)). The bandgap of rGO–TiO_2 is narrower than that of TiO_2 composites, which is in favor of the absorption of visible light. The second reason is that rGO has good conductivity and can easily transfer photogenerated electrons to the electrode. This conclusion is also proved by the impedance spectrum (Figure 6.24), where the charge-transfer resistance of the composite decreases from 417 Ω cm^2 of TiO_2 to 209 Ω cm^2. The rapid electron transfer between TiO_2 and rGO will greatly reduce the recombination rate of electrons and holes. By introducing PANI into the composites, the blank photocurrent sharply increased to 110.8 nA. As a kind of conducting polymer, PANI can further increase the conductivity of the composite (curve 3 in Figure 6.24). At the same time, it also has high absorptivity and carrier transfer rate, which endow PANI excellent photoelectric property of the composite materials.

When 166.8 μmol/L GA was added to the photoelectron system, the photocurrent of rGO/TiO_2/PANI-modified electrode rapidly increased to 199.8 nA, which was 5.9 and 1.2 times that of TiO_2 and rGO–TiO_2, respectively. The increase in the photocurrent of the three materials is mainly attributed to the fact that the photogenerated holes can oxidize GA, and the holes can be further excited by the electrons. Compared with the blank photocurrent, the photocurrent increased by 79% after the addition of GA, which is enough to show that rGO/TiO_2/PANI can be used as a good photoelectric material for the detection of GA antioxidant capacity.

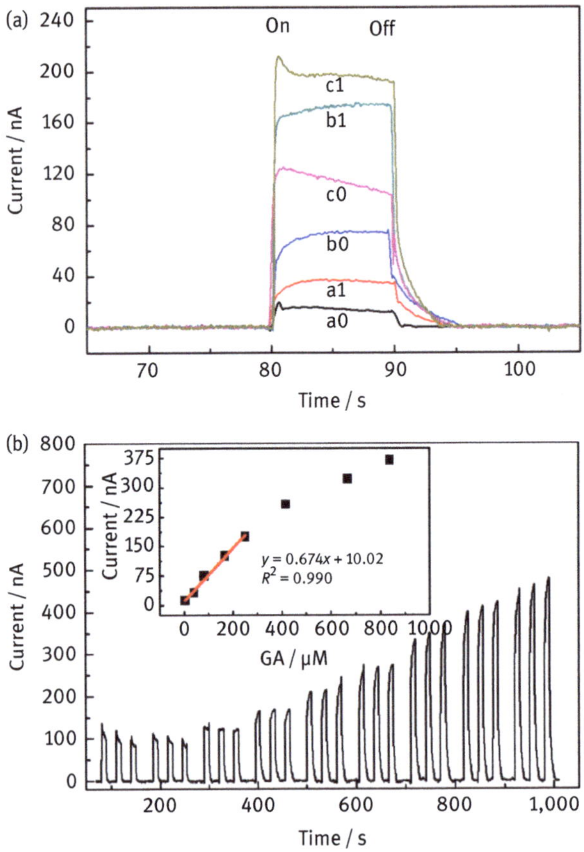

Figure 6.22: (A) Photocurrent response of ITO (indium tin oxides) electrode modified by TiO_2 (a_0, a_1), rGO–TiO_2 (b_0, b_1) and rGO/TiO_2/PANI (c_0, c_1) (a_0, b_0 and c_0 are blank PBS (phosphate buffer saline); a_1, b_1 and c_1 are 166.8 μmol/L GA for the sample concentration). (B) Photocurrents corresponding to different concentrations of GA (reaction conditions: voltage, 0 V; light source, LED 420 nm; buffer solution, 0.1 mol/L; pH = 7.4; illustration for the GA's linear curve).

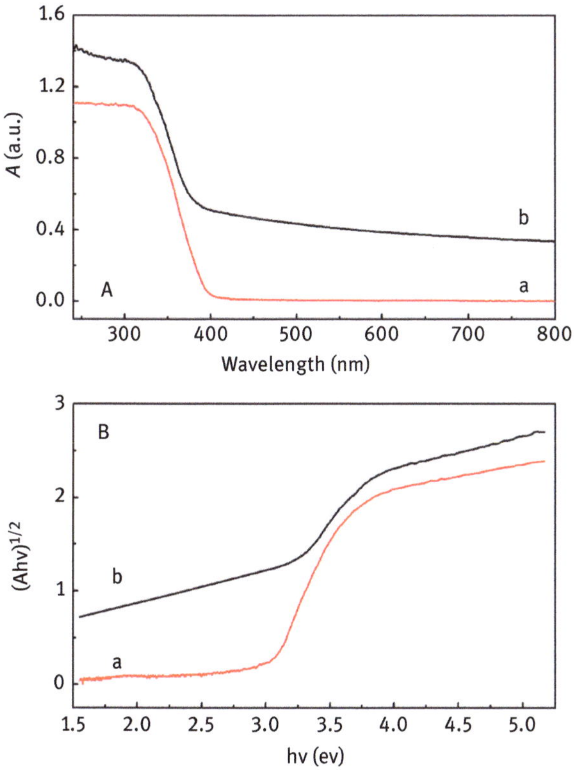

Figure 6.23: (A) Diffuse reflections of rGO–TiO_2 (a) and TiO_2 (b); (B) Kubelka–Munk transformation of (A).

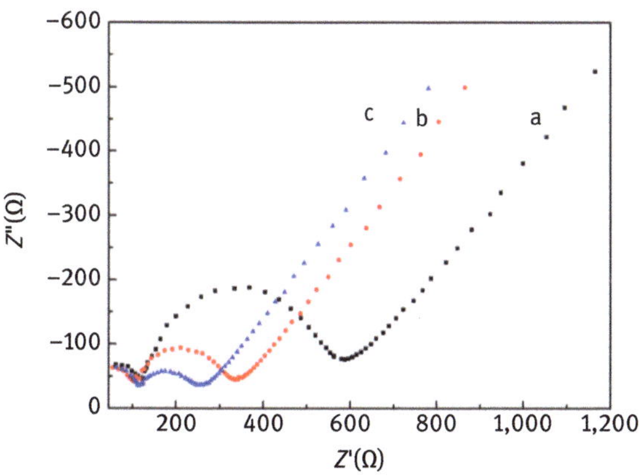

Figure 6.24: Impedance of TiO_2 (1), rGO–TiO_2 (2) and rGO/TiO_2/PANI (3) in 1 mmol/L [Fe $(CN)_6]^{3-}/^{4-}$ and 0.1 mol/Atlas (amplitude, 10 mV; frequency, 10^{-1}–10^5 Hz).

6.4.1.3 Optimization of experimental conditions

First, the content of PANI in rGO/TiO_2/PANI composites was optimized. The researchers synthesized rGO/TiO_2/PANI with the mass ratios of PANI and rGO–TiO_2 of 0.75%, 1.5%, 3%, 6% and 12% composites. The corresponding photocurrent was measured. As shown in Figure 6.25, first, the photocurrent increased and then decreased with the increase in the PANI content; hence, the highest photocurrent was at 1.5%. The excellent properties of PANI increased the photocurrent, but a more amount of PANI inhibits the transfer of electrons from PANI LUMO (lowest unoccupied molecular orbital) to TiO_2 CB (conduction band). Therefore, for further analysis, only 1.5% PANI is used.

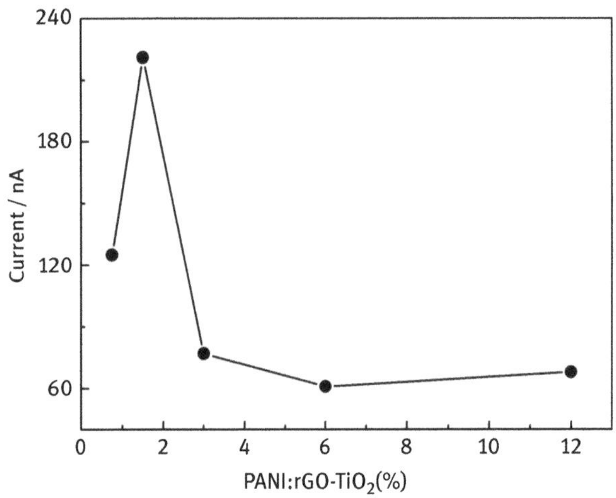

Figure 6.25: Effect of different ratios of PANI on photocurrent in composite materials (reaction conditions: GA concentration, 166.8 μmol/L; voltage, 0 V; light source, LED of 420 nm; buffer solution, 0.1 mol/L; pH = 7.4).

The applied potential in photoelectrochemical sensing is a very important factor. As shown in Figure 6.26, the photocurrent first increases and then reaches a plateau within the applied potential range of −0.1 to 0.3 V. For the n-type semiconductor, initially there is a very large gradient of electrons between the electro-optical material and the ITO electrode. When the positive potential is applied, the electrons flow from the conduction band of the photoelectric material to the ITO electrode. The more positive the applied potential, the more conducive to the transfer of electrons to the ITO electrode to produce greater photocurrent. When the electron gradient between the photoelectron material and ITO reaches the maximum (0.1 V), the photocurrent also reaches the maximum. When the potential is increased sequentially, the photocurrent does not increase. When the applied potential is 0 V, the corresponding photocurrent is 56.3% of the maximum photocurrent, which already presents a high sensitivity. Thus, 0 V is selected as the measured potential for the photoelectrochemical sensor.

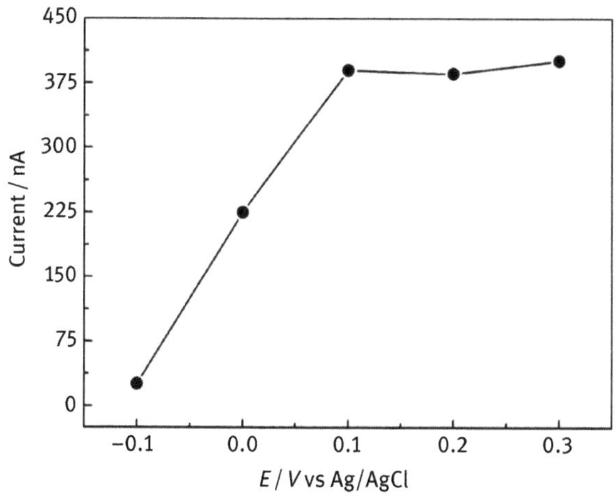

Figure 6.26: Effect of different potentials on photocurrent (reaction conditions: GA concentration, 166.8 μmol/L; voltage of 0 V; light source, LED 420 nm; buffer solution, 0.1 mol/L; pH = 7.4).

6.4.1.4 Determination of GA selectivity

To determine GA (Figure 6.27), first, the sample was added to the photoelectric test cell. In order to prevent the interference of colored materials, the visible light is irradiated from the back of the modified electrode, and finally the photoelectric signal is collected. When the wine sample is measured, it is not treated directly. To measure samples, the photocurrents are collected using the following rules: $I = I_{sample} - I_{blank}$ (I_{blank} is the photocurrent when no sample is added, and I_{sample} is the photocurrent after the addition of the sample).

Under the optimum conditions, the antioxidant capacity of GA was determined using a photoelectric platform prepared by rGO/TiO_2/PANI composite. As shown in Figure 6.22(b), in the range of 4.17–838.00 μmol/L, the photocurrent increased with the increase in the concentration of GA, showing a good linearity in the range of 4.17–250.02 μmol/L, and the lowest detection limit is 1.72 μmol/L. The relative standard deviation (RSD) of the six electrodes was 3.6%. Once used, the modified electrode was stored at room temperature for 4 weeks, and the measured signal still retained 90.3% of the original signal. The designed photoelectric sensor can be well used in the actual sample detection and analysis because of its wide detection range, low detection limit, long time stability and fast response.

The researchers also performed anti-interference assays on reducing substances (GSH and Cys) that are present for determination. The experimental results are shown in Figure 6.28. About 200-fold GSH and Cys do not affect the determination

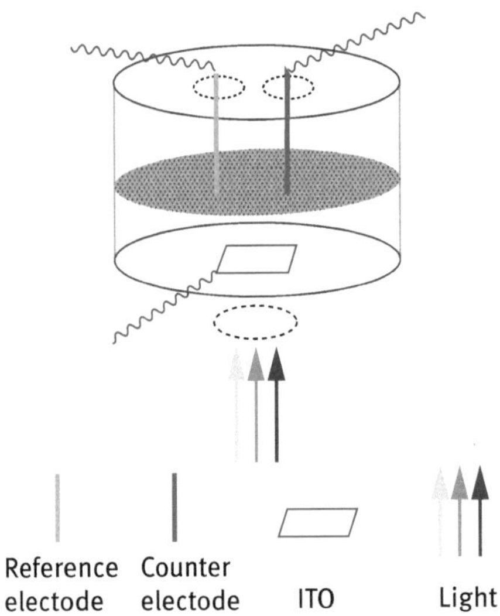

Figure 6.27: Scheme of antioxidant capacity of a photodetection cell.

of GA. Compared with GSH and Cys (0.533 V, Figure 6.28 (b)), GA showed a lower oxidation potential of 0.019 V (Figure 6.28 (c)); hence, photogenerated cavities first selected GA to react and at a certain concentration, GSH and Cys do not interfere with the assay. AA and GA showed similar redox potentials, but the same concentration of

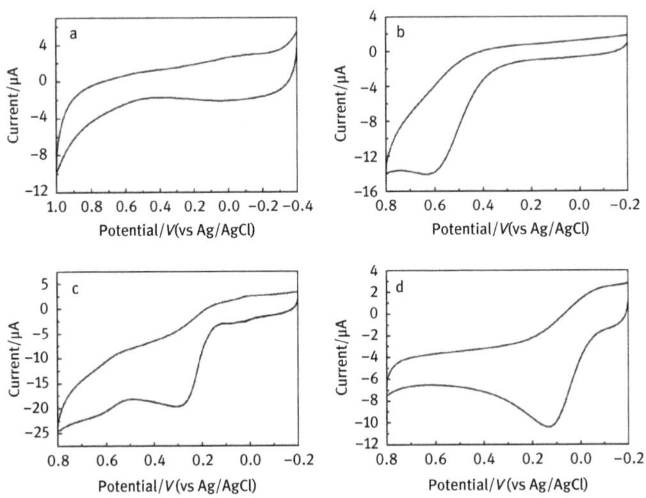

Figure 6.28: Cyclic voltammograms of 1.25 mmol/L GSH (a), Cys (b), GA (c) and AA (d) in 0.1 mol/L PBS at a scan rate of 100 mV/s.

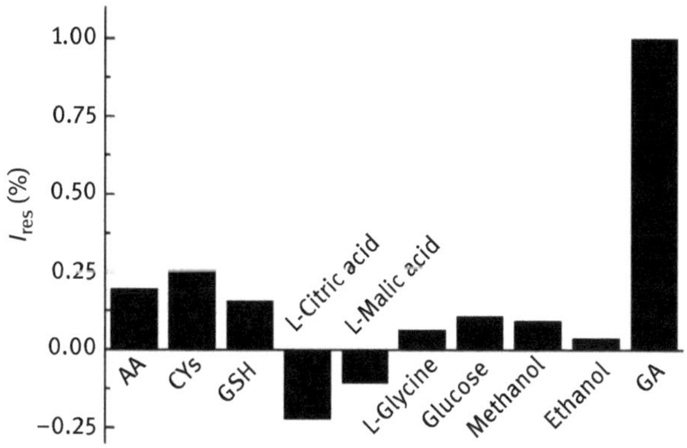

Figure 6.29: Photoelectrochemical determination of various interfering substances (concentration of test substance: 166.8 μmol/L AA and GA; 33.36 mmol/L GSH, Cys, L-citric acid and L-malic acid; 0.167 mol/L ethanol, methanol, dextrose and L-glycine. Buffer solution: 0.1 mol/L PBS, pH = 7.4; operating potential 0 V; LED light source, 420 nm).

AA did not interfere with the determination of GA (Figure 6.29). However, AA showed a strong photoelectric response at the rGO–TiO_2-modified electrode. Therefore, the addition of PANI increased the selectivity of the photosensor built with rGO/TiO_2/PANI. Meanwhile, we measured the other interfering substances: 1,000 times of the methanol, ethanol, glucose and glycine and 200 times of the citric acid and malic acid do not interfere with the determination of GA.

Using the designed photoelectric sensor, researchers measured the four different brands of red wine and the results are shown in Table 6.1. All four wines contain high GA and the result of the photoelectrochemical determination is generally lower compared with the F-C method, because the F-C method determines the content of the total phenols in the wine while the photoelectric method determines the content of the GA. The results are consistent with the HPLC results, which further proved the selective determination of GA by photoelectric sensors.

Table 6.1: Antioxidant capacity assay of the red wine by different techniques.

Red wine	Photoelectrochemical sensor	F-C method	HPLC
Brand 1	692.4 ±3.5	713.8 ±2.6	427.3 ±2.3
Brand 2	1,083.5 ±5.3	1,244.3 ±3.2	769.4 ±4.5
Brand 3	313.2 ±1.4	843.6 ±2.3	286.6 ±1.6
Brand 4	166.7 ±1.1	663.5 ±1.8	143.9 ±0.7

6.4.1.5 Reaction mechanism of antioxidant capacity by photoelectric analysis

In elaborating on the mechanism of designing photosensors, Hu thought that the hydroxyl radicals generated by hydroxyl groups or water adsorbed on semiconductors participated in the reaction of measuring reactants, but he did not give any direct evidence for the existence of hydroxyl radicals. When *p*-dibenzoic acid reacted with the hydroxyl radical, the resultant fluoresced at 315 nm. Based on this principle, hydroxyl radicals can be detected. As shown in Figure 6.30, no fluorescence was detected in the solution in the presence of rGO/TiO_2/PANI, when visible light irradiates for 2 min,; therefore, hydroxyl radicals did not participate in the photoelectric sensing reaction of rGO/TiO_2/PANI. The possible mechanism of PANI/rGO/TiO_2 photoelectric sensor is shown in Figure 6.31. Under visible light irradiation, electrons transition from HOMO of PANI to LUMO and then from the valence band to the conduction band as the doped TiO_2 is excited. From the results of diffuse reflectance in Figure 6.32(a) and Figure 6.23(b), the conduction band of rGO–TiO_2 is at 0.4 V (vs Ag/AgCl), which is more positive than that of PANI at LUMO. Therefore, electrons flow from the LUMO of PANI to the conduction band of rGO–TiO_2, and then flow into the ITO electrode to generate photocurrent. At the same time, the good conductivity of graphene and PANI accelerate the transfer of electrons, so that the sensor shows a rapid response. When GA is added, electrons are transferred from the valence band of rGO–TiO_2 to the HOMO holes of PANI and the holes generated by PANI themselves react with GA, and the holes of the electrons can be further excited by the photocurrent amplification. At a certain range, the higher the GA concentration, the higher the photocurrent.

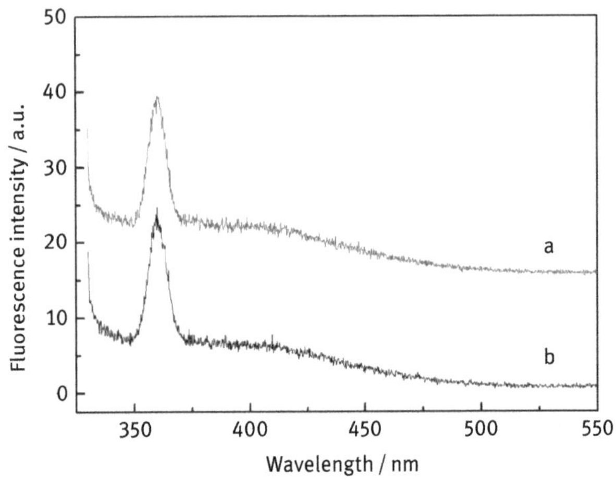

Figure 6.30: Fluorescence emission spectra of 0.5 mmol/L terephthalic acid in the presence of rGO/TiO_2/PANI catalyst at 2 min light (a) and no light (b).

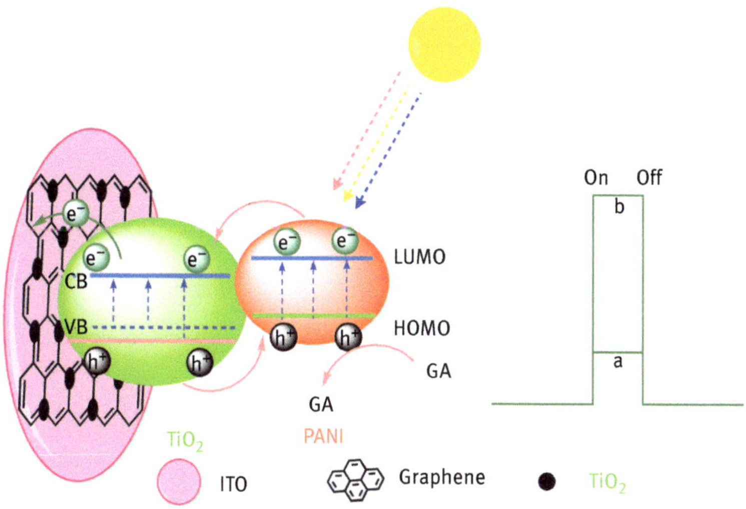

Figure 6.31: Mechanism diagram of photoelectrochemical detection of GA: (a) light and (b) no light.

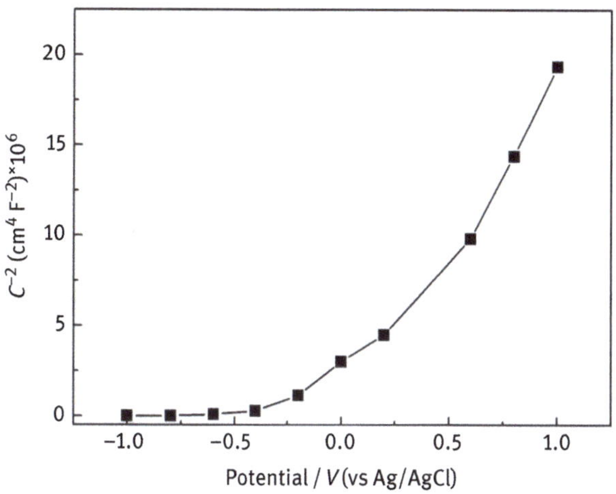

Figure 6.32: The Mott–Schottky plot of rGO–TiO_2-modified electrode (at 1,000 Hz) in 0.1 mol/L PBS (pH 7.4).

Using the nanocomposite rGO/TiO_2/PANI as a photoelectric material, a new type of photoelectric sensor device has been designed, successfully realizing the selective determination of antioxidant capacity of GA in red wine. When compared with the traditional method, photoelectrochemical sensing shows the characteristics of fast response, high sensitivity and good selectivity. Therefore, this chapter provides a new method for the determination of antioxidant capacity of GA in foods and beverages, and it also provides a good foundation for the fingerprint analysis of antioxidant capacity in foods.

6.4.2 Determination of total antioxidant capacity using photoelectrochemical platform

The selection of optoelectronic materials in the photoelectrochemical platform is of crucial importance. TiO_2 has drawn much attention because of its excellent photostability, chemical and biological inertness, nontoxicity and high carrier concentration. However, its low photoelectric efficiency and ability to absorb only ultraviolet light greatly limit its application. Inorganic semiconductors such as CdS, ZnS and MoS_2 are often used for the modification of TiO_2 because of their absorption of visible light, but their toxicity also limits further application. Recently, g-C_3N_4 has been widely used in photocatalytic and photoelectrochemical studies because it has no metal elements, is nontoxic and has a narrow bandgap, and its various morphologies are also easy to prepare. As the energy band positions of g-C_3N_4 and TiO_2 match well and achieve rapid separation of photogenerated electrons and holes to greatly improve the photoelectric efficiency, some scholars have used it to improve the performance of TiO_2 and used the composite in hydrogen evolution and environmental purification. However, the low specific surface area and poor solubility of bulk g-C_3N_4 limit its further application. Hence, researchers used different methods to prepare g-C_3N_4 with thin layer for improving its surface area and solubility. Therefore, g-C_3N_4 with high specific surface area can provide more active sites, and its photoelectric properties when compared with bulk g-C_3N_4 showed an unparalleled advantage, which recent theoretical studies have also validated. Based on this, researchers synthesized ultrathin g-C_3N_4 (utg-C_3N_4) and utg-C_3N_4/TiO_2 composites using utg-C_3N_4 and $TiCl_3$ as starting materials. Then a thin-film photovoltaic cell was designed using utg-C_3N_4/TiO_2 as a photovoltaic material, and successfully used in the flow measurement of the antioxidant capacity of total antioxidants [32]. The system exhibits ease of operation, rapid response, resistance to contamination and interference with colored objects.

Figure 6.33 shows the flowchart for determining the total antioxidant capacity of antioxidants using utg-C_3N_4/TiO_2 as a photovoltaic material. The bulk g-C_3N_4 was synthesized by thermal polymerization and g-C_3N_4 with thin layer was obtained by a simple sonication method. Then, the utg-C_3N_4/TiO_2 nanocomposite was synthesized by the hydrothermal method. Finally, using the composite as a photovoltaic material, the total antioxidant capacity of the antioxidant was determined in a thin-layer flow cell.

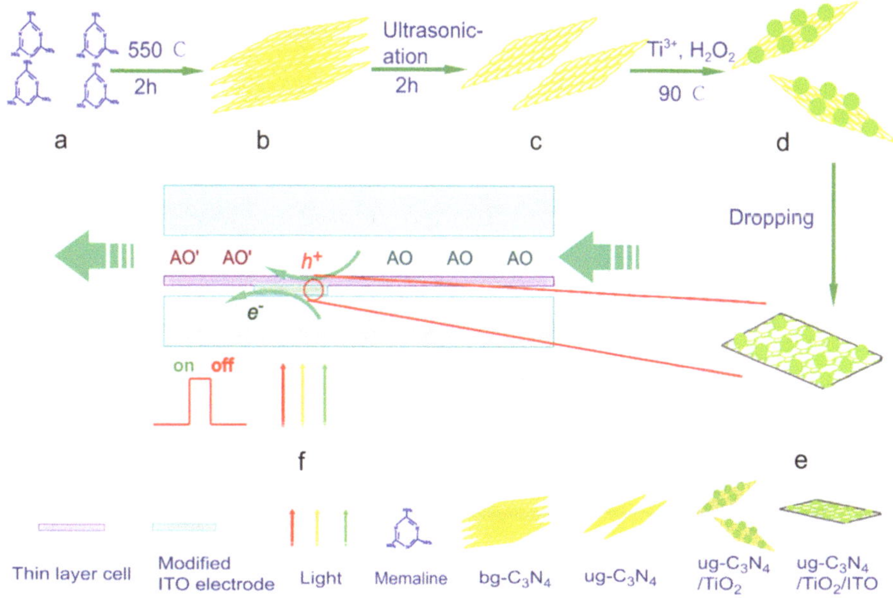

Figure 6.33: Flowchart of global antioxidant capacity evaluation: utg-C_3N_4/TiO_2 synthesis process (a–e) and photodetection of thin-layer diagram (f). AO, antioxidant; AO', antioxidant after oxidation.

6.4.2.1 Structure and morphology of materials

First, the synthesized material utg-C_3N_4 was characterized by atomic force microscopy (AFM). As shown in Figure 6.34(a), the thickness of the utg-C_3N_4 layer was measured at 1.2 nm (Figure 6.34(b)), which indicates that bg-C_3N_4 has been successfully peeled into thin layers. Transmission electron microscopy (TEM) showed results similar to AFM, and utg-C_3N_4 was more transparent than bg-C_3N_4 (Figure 6.34 (c), (d)). As shown in Figure 6.35, no precipitation of utg-C_3N_4 aqueous solution occurs in a few weeks, while bg-C_3N_4 is deposited on the bottom, indicating that utg-C_3N_4 has better solubility. TiO_2 nanoparticles about 10 nm are uniformly distributed on the utg-C_3N_4 layer (Figure 6.34 (e)), while the bulk C_3N_4 is clearly aggregated (Figure 6.35 (c)). The high-resolution image clearly shows that the lattice spacing of TiO_2 is 0.35 nm (Figure 6.34 (f)), which is an indirect reflection of the TiO_2 (101) crystal plane, indicating that the crystalline form of TiO_2 is anatase. This is also confirmed by the results of high-resolution FFT (fast Fourier transformation) conversion (Figure 6.34 (f)).

6.4.2.2 Determination of total antioxidant capacity of antioxidants

In the photoelectric measurement process, first, mobile photoelectrochemical platform is fixed. Then the PBS and the added PBS solution flow through the sample pool at a speed of 2 rev/min under the assistance of the flow pump. In order to

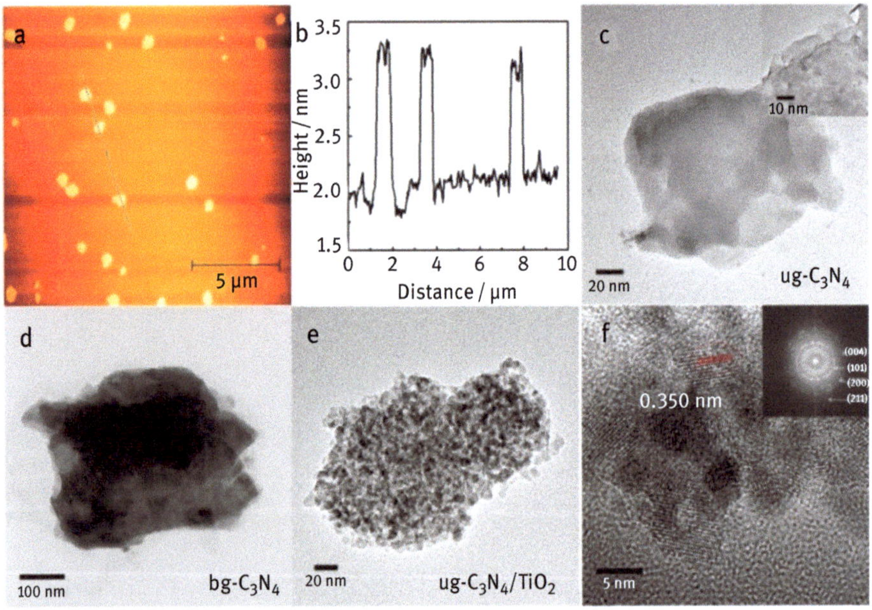

Figure 6.34: AFM images (a) and corresponding height map (b) of utg-C_3N_4; TEM image of utg-C_3N_4 (c); TEM image of bg-C_3N_4 (d); TEM image of utg-C_3N_4/TiO_2 and HRTEM image (f). Interpolation (c) is the HRTEM plot of utg-C_3N_4 and the interpolation plot (f) is the FFT of the utg-C_3N_4/TiO_2 HRTEM plot.

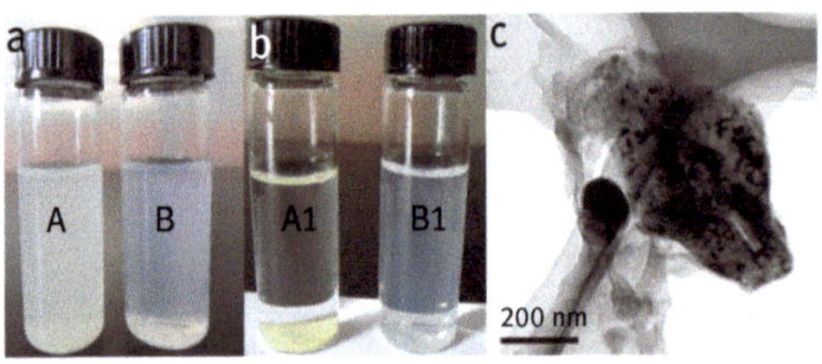

Figure 6.35: TEM image (a) of bg-C_3N_4 (A, A1), utg-C_3N_4 (B, B1) (A1 and B1 are two weeks later (b)) and bg-C_3N_4/TiO_2 (c).

avoid color interference, the LED light source is irradiated from the modified ITO electrode, and the energy of the electrode surface is 73.89 mW/cm^2. Each sample was measured three times to get the average. When the sample is measured, the photoelectric flow follows the following rules to be collected: $I = I_{\text{sample}} - I_{\text{blank}}$ (I_{blank} is photocurrent without sample addition and I_{sample} is photocurrent after the addition of sample).

Photoelectrochemical measurement has the advantages of both optics and electrochemistry, which has attracted the attention of researchers widely. In the study of photoelectrochemistry, photoelectric materials are the key. The photocurrent produced by TiO_2 nanoparticles is only 13.8 nA because it only has weak absorption of visible light. When TiO_2 and bg-C_3N_4 are combined, the photocurrent increases rapidly to 76.85 nA, which is mainly due to the great increase in the absorption of visible light by bg-C_3N_4 (Figure 6.36). To the surprise of researchers, the photocurrent sharply increased to 1.7 times that of bg-C_3N_4/TiO_2 when composited with utg-C_3N_4 and TiO_2. The good photoclcctric properties of utg-C_3N_4/TiO_2 are mainly attributed to three reasons: (1) same as bg-C_3N_4, utg-C_3N_4 has good absorption of visible light (Figure 6.36); (2) compared with bg-C_3N_4, utg-C_3N_4 has good electrical conductivity, which can be confirmed by impedance spectroscopy (Figure 6.37 (b)); and (3) compared with the agglomeration of TiO_2 nanoparticles on the surface of bg-C_3N_4, utg-C_3N_4 disperses the TiO_2 nanoparticles on the surface. The good contact between TiO_2 nanoparticles and utg-C_3N_4 surface will facilitate the transfer of photogenerated electrons and reduce the compound of electrons and holes, thereby improving the optoelectronic properties. As shown in Figure 6.37 (c), the charge transfer resistance (R_{ct}) of utg-C_3N_4/TiO_2 is 658.6 Ω cm^2 under visible light irradiation while the R_{ct} of bg-C_3N_4/TiO_2 is 1,429.0 Ω cm^2, demonstrating that utg-C_3N_4/TiO_2 has the good ability of electron mass transfer. To better illustrate this problem, fluorescence quenching experiments were performed. As shown in Figure 6.37 (d), the fluorescence of utg-C_3N_4 decreases at 420 nm with the addition of TiO_2 nanoparticles. This is mainly because the suitable bandgap positions of utg-C_3N_4 and TiO_2 facilitate the separation of electrons and holes and the compound reduced. When 50 μmol/L CT was added into the system, the photocurrent of the three modified electrode chips increased rapidly and that of utg-C_3N_4/TiO_2-modified electrode increased to 284.28 nA, which was 1.7 and 5.1 times to that of bg-C_3N_4/TiO_2 and TiO_2-modified electrode, respectively. A similar phenomenon was observed with the addition of other antioxidants. This fully shows

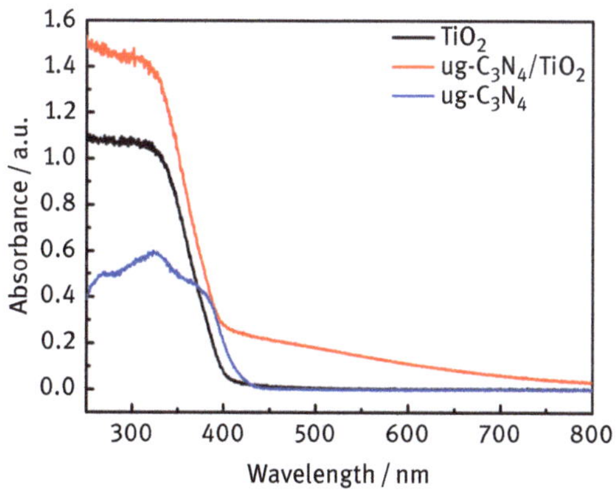

Figure 6.36: Diffuse reflectance absorption of TiO_2, utg-C_3N_4 and TiO_2/utg-C_3N_4.

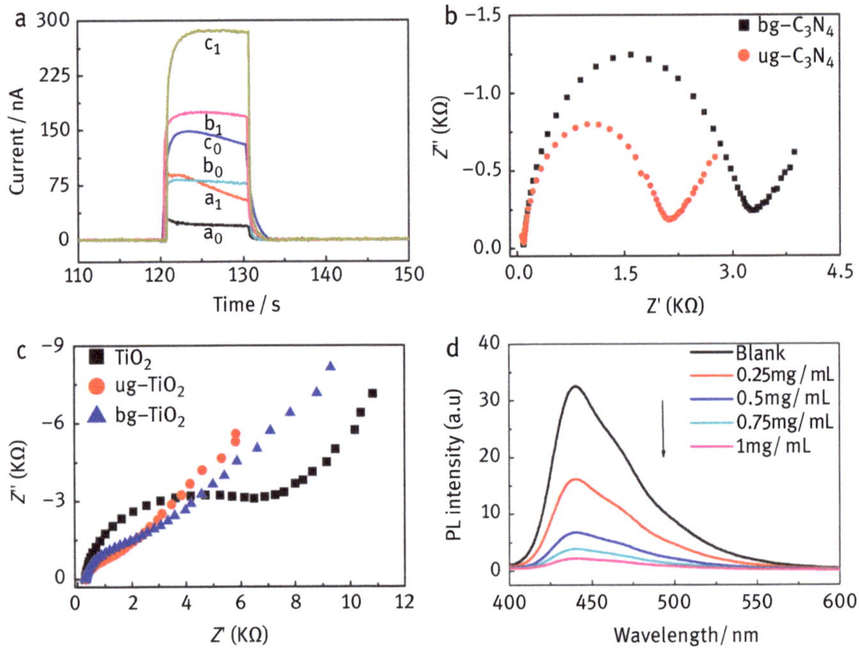

Figure 6.37: Photocurrent response of ITO electrode modified with TiO_2 (a_0, a_1), bg-C_3N_4/TiO_2 (b_0, b_1) and utg-C_3N_4/TiO_2 (c_0, c_1) in blank solution (a_0, b_0, c_0) and 50 μmol/(a_1, b_1, c_1) (a); electrochemical impedance spectroscopy of bg-C_3N_4, utg-C_3N_4 (b) and TiO_2, bg-C_3N_4/TiO_2, utg-C_3N_4/TiO_2 (c); fluorescence emission of utg-C_3N_4 at different TiO_2 concentrations (d). The glassy carbon electrode was used as the substrate electrode for electrochemical impedance spectroscopy (EIS) in a mixed solution of 5 mmol/L $[Fe(CN)_6]^{3-/4-}$ and 0.1 mol/L KCl in (b). (c) The ITO electrode was used as the substrate electrode for modification under the condition of 0.3 V versus Ag/AgCl at 420 nm, and the electrochemical impedance spectroscopy was carried out in the mixed solution of 0.6 mol/L Na_2SO_4.

that utg-C_3N_4/TiO_2 can be used for antioxidant capacity evaluation of antioxidants as high-performance optoelectronic materials.

In the photoelectrochemical sensing method, the potential is a very important factor. As shown in Figure 6.38, the photocurrent first increases and reaches a plateau within −0.07 to 0.3 V applied potential range. For the n-type semiconductor, initially there is a concentration difference between the electro-optical material and the ITO electrode. When a positive potential is applied, the electrons flow from the conduction band of the photoelectric material to the ITO electrode. The more the applied potential is, the more positive the electrons are transferred to the ITO electrode to generate a larger photocurrent. When the gradient of the electron between the photoelectric material and the ITO electrode reaches the maximum (0 V), the photocurrent reaches the maximum, and then the photocurrent will no longer increase significantly. Therefore, 0 V is selected for the photoelectric sensor's measured potential.

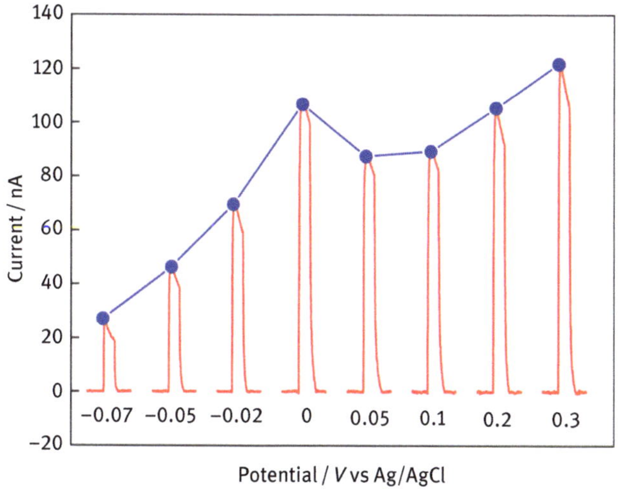

Figure 6.38: The curve of the change of the photocurrent with the potential.

At 0 V, researchers first measured the antioxidant capacity of nine standard antioxidants using a thin-film photovoltaic cell, and all nine antioxidants exhibited good photoresponse. The photoelastic response of CA is shown in Figure 6.39, with a linear range of 15.00–193.31 and an RSD of 4.6%. The results of the other eight antioxidants are shown in Table 6.2. The use of utg-C_3N_4/TiO_2 for 2 weeks still retains 95% of the initial signal, indicating that thin-film photovoltaic cells have better resistance to poisoning of electrodes.

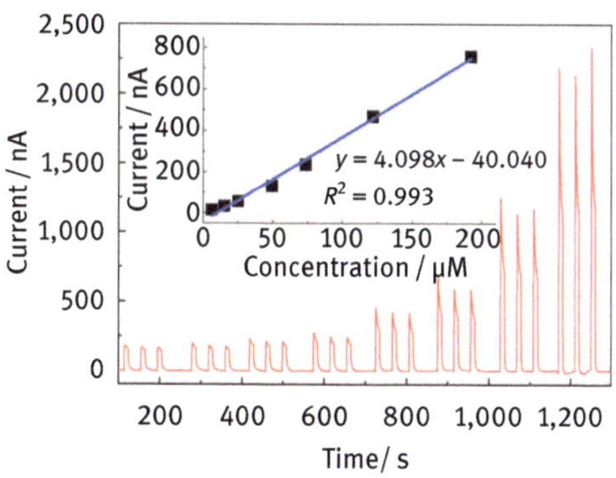

Figure 6.39: Photoelectric response of CA on optoelectronic devices and its linear equation. Applied voltage is 0.0 V and the wavelength of light is 420 nm.

Table 6.2: Linear equation, correlation coefficient (R^2), redox potential, ion potential, antioxidant capacity and determination of concentration range of nine antioxidants.

Antioxidants	Linear equation	Linearity range (μmol L^{-1})	R^2	E (V)	IP (kal mol^{-1})	Antioxidant capacity
Quercetin	$y=3.150x-14.62$	9.90–111.10	0.992	0.300	161.89	3.150
Gallic acid	$y=3.17x+12.21$	5.00–74.44	0.956	0.217	179.65	3.170
Caffeic acid	$y=4.099x-40.04$	15.00–193.31	0.993	0.374	175.72	4.099
Catechin	$y=5.195x-79.82$	24.96–192.31	0.992	0.394	166.27	5.195
Fisetin	$y=3.375x-89.35$	29.56–181.82	0.983	0.189	266.27	3.375
Rutin	$y=1.571x+14.175$	24.88–192.31	0.973	0.447	165.12	1.571
Trolox	$y=0.117x+25.31$	73.85–348.84	0.997	0.356	−6.73	0.117
Ascorbic acid	$y=0.629x+46.28$	25.00–243.09	0.970	0.365	182.64	0.629
Glutathione	$y=0.039x+8.824$	9.65–400.00	0.996	–	179.64	0.039

6.4.2.3 The principle of total antioxidant capacity by photoelectric analysis of antioxidants

The two widely accepted principles of free radical and antioxidant reaction in the field of antioxidation are: (1) hydrogen atom transfer mechanism and (2) single electron transfer mechanism. As shown in Figure 6.40 (a), the valence band of utg-C3N4 is found at −6.257 eV by the Mott–Schottky and diffuse reflectance mapping, which is lower than the water's cleavage potential (−5.30 eV); hence, in theory it can oxidize water molecules. The oxidation of water molecules is often accompanied by the generation of hydroxyl radicals. When *para*-dibenzoic acid was added to the aqueous solution containing utg-C_3N_4/TiO_2 catalyst, no fluorescence was observed at 425 nm after different times of irradiation (Figure 6.40 (b), which indicates

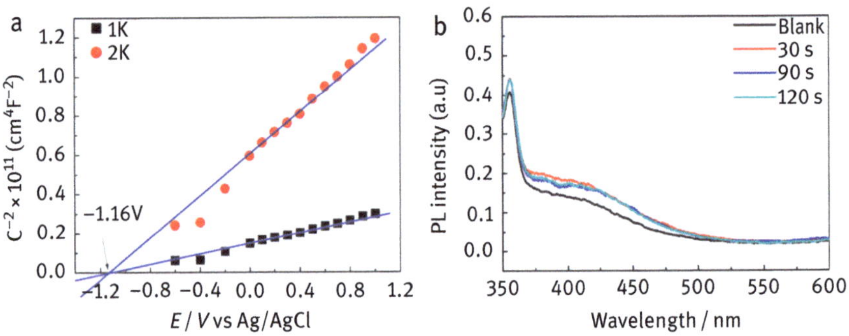

Figure 6.40: The Mott–Schottky plot of utg-C_3N_4 in 1 mol/L Na_2SO_4 (a); fluorescence emission spectra of 0.5 mmol/L dibenzoic acid in the presence of utg-C_3N_4/TiO_2 at 0, 30, 90 and 120 s, respectively (b). (a) The frequencies are 1,000 and 2,000 Hz and the applied potential range is −0.6 to 1.2 V; and (b) the excitation light is 315 nm.

that hydroxyl radicals were not generated in the system. Therefore, the principle of photoelectrochemical platform based on utg-C_3N_4/TiO_2 catalyst can be considered as a single electron transfer mechanism.

As shown in Figure 6.41, under visible light irradiation, the electron transitions from the valence band of utg-C_3N_4 to the conduction band, while the valence band leaves holes to be generated. Because the position of the conduction band between utg-C_3N_4 and TiO_2 matches well, electrons rapidly reach the ITO electrode through the conduction band of TiO_2 to generate photocurrent. When an antioxidant is added to the system, the antioxidant reacts with holes in the valence band of utg-C_3N_4 to regain the electrons for further excitation, which amplifies the photocurrent signal. Therefore, based on the one-electron transfer in the principles discussed earlier, the antioxidant capacity can be determined rather than the concentration of antioxidants.

In Figure 6.42, CT, CA, FT, GA, QR and RT shows strong antioxidant capacity compared to TR, AA and GSH. The earlier analysis shows that the photoelectrochemical platform based on utg-C_3N_4/TiO_2 catalyst is carried out by the single electron transfer mechanism. The stronger the antioxidant capacity, the stronger the ability to

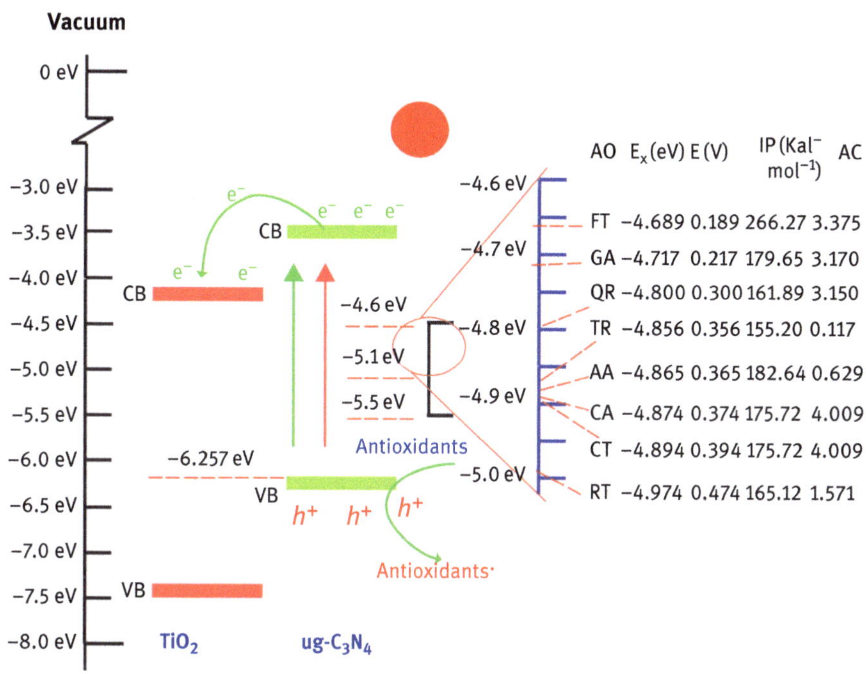

Figure 6.41: The mechanism map for detecting the antioxidant capacity of photochemical antioxidants. AO, antioxidant; E_x (eV), relative vacuum redox potential of antioxidant; *E* (V), redox potential of antioxidant to Ag/AgCl; IP (kcal/mol): ion potential of antioxidants; antioxidant capacity is derived from the straight line equation of antioxidants.

lose the electron. To study the reasons for different antioxidants with different antioxidant capacities of distinct species, the researchers also conducted both theoretical and experimental exploration. The ionic potential of antioxidants is an important indicator for the antioxidant capacity of antioxidants. The ionic potential of nine antioxidants was calculated using the B3LYP/6–311G (d, p) method. The results are shown in Figure 6.41. The ionic potential of the phenolic catalysts CT, CA, FT and GA is low, which corresponds to their strong antioxidant capacity. This is also demonstrated by comparing the ionization potential of cysteine (187.95 kcal/mol) and glucose (191.17 kcal/mol), because both have a high ionic potential and do not readily react without affecting the actual sample detection. However, GA and TR are somewhat inconsistent with the above inference. Redox potential is another important indicator of antioxidant capacity; the lower the redox potential, the stronger the antioxidant capacity. As shown in Figure 6.42, GA possesses a low redox potential while exhibiting a relatively high antioxidant capacity. Other phenolic antioxidants also show lower redox potential and higher antioxidant capacity than GSH. In addition, it should be noted that the molecular structure plays an important role in the photoelectrochemical reaction. Theoretically, the position and number of hydroxyl groups on the phenolic benzene ring strongly affect the antioxidant capacity of phenols. Compared with TR, CT, FT, QR and RT benzene ring has two *ortho*-hydroxyl groups, showing a strong antioxidant capacity. The ease of contacting antioxidants with the catalyst is another important factor that determines the degree of antioxidant reactions. Since utg-C_3N_4 has a benzene ring-like structure, phenols having a benzene ring are more likely to react with the catalyst through the π–π interaction and thus show strong antioxidant capacity. In summary, compared with TR, AA and GSH in particular, phenolic antioxidants show strong antioxidant capacity, which is also consistent with the optical determination results of other researchers.

6.4.2.4 The photoelectric analysis determination of the antioxidant capacity of actual sample

In order to test the actual sample, the researchers first measured the interfering substances commonly found in food. The results are shown in Figure 6.43: 1,000 times of the amount of proline, glycine, histidine, ethanol and methanol; 500 times of the amount of threonine, fructose, glucose, citric acid and malic acid; and 20 times of the amount of cysteine have no effect on the photoelectric analysis in the presence of 25 μmol/amino acids. This further demonstrates that the optoelectronic platform can be used for the determination of complex real samples. The antioxidant capacity of four brands of coffee and tea purchased on the market is tested. First, four commercial tea and coffee samples were treated as follows: For each tea sample weighing 0.5 g, add 50 mL of boiling water and filter it after 30 min. The obtained filtrate was directly measured. The results shown in Table 6.3 are in good agreement with those of the two optical methods. The results of the four kinds of

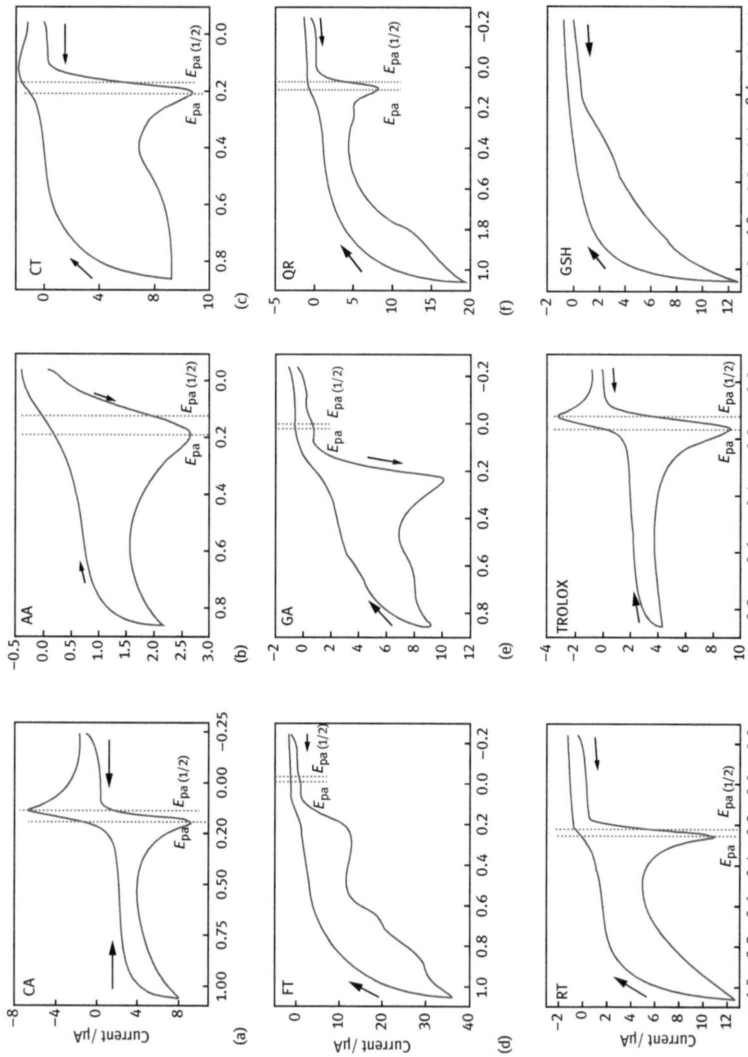

Figure 6.42: Cyclic voltammogram of nine antioxidants (antioxidant concentration is 0.5 mm/L and potential scan rate is 100 mV/s).

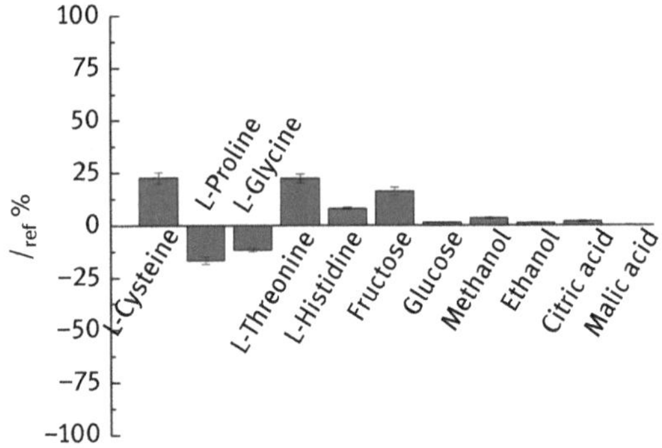

Figure 6.43: About 25 μmol/L of GA is used as a sample. The results showing the influence of common interfering substances in foods when measuring the antioxidant capacity of antioxidants are 25 mmol/L L-proline, L-glycine, L-histidine, ethanol, methanol; 12.5 mmol/L L-threonine, fructose, glucose, L-citric acid; 0.5 mmol/L L-cysteine in 0.1 mol/L PBS (pH = 7.4) with 0 V potential and 420 nm LED illumination.

Table 6.3: Comparison of test results of three different methods on four brands of tea and four brands of coffee.

Practical samples	Photoelectrochemical sensor ($mg\bullet g^{-1}$ GA)	Folin–Ciocalteu method ($mg\bullet g^{-1}$ GA)	DPPH method ($mg\bullet g^{-1}$ TROLOX)
T_1	72.212 ± 0.612	62.190 ± 0.426	163.610 ± 0.925
T_2	95.860 ± 0.536	88.610 ± 0.731	227.880 ± 0.834
T_3	21.010 ± 0.214	23.270 ± 0.198	113.600 ± 0.736
T_4	30.570 ± 0.165	34.960 ± 0.767	132.560 ± 0.961
C_1	3.334 ± 0.068	9.020 ± 0.213	14.740 ± 0.346
C_2	3.759 ± 0.086	9.645 ± 0.324	15.490 ± 0.425
C_3	1.937 ± 0.034	5.520 ± 0.228	6.890 ± 0.123
C_4	2.353 ± 0.056	6.830 ± 0.621	9.682 ± 0.456

coffee are generally lower than those of the Folin–Ciocalteu (F-C) method, which is due to the interference of the coffee color. The photoelectrochemical thin-film flow platform illuminates the back of the ITO-modified electrode, which overcomes the problem of interference of the colored sample well and makes the determination result more accurate.

Photoelectrochemical method was first introduced as a new measurement method in the field of antioxidant capacity. This method overcomes the three main drawbacks of traditional methods: (1) chromatographic methods only measure the concentration of antioxidants rather than the capacity, and in particular, cannot reflect the synergy between antioxidants; (2) spectral methods are susceptible to the interference of colored substances; and (3) the product in traditional electrochemical methods easily adsorbs on the electrode and affects further determinations. Advantages of photoelectrochemical methods are rapid response, high sensitivity, good stability and less injection. At the same time, it is proved by experiments and theoretical analysis that the photoelectrochemical detection is a single electron transfer mechanism. The measured results are in accordance with the reference spectrum method. In view of these significant advantages, the photoelectrochemical platform is expected to become a versatile one for determining the antioxidant capacity of antioxidants in foods.

References

[1] Linsebigler A L, Lu G Q, Yates J T. Photocatalysis on TiO2 Surfaces: Principles, Mechanisms, and Selected Results. Chemical Reviews, 1995, 95: 735–758.
[2] Chen X B, Shen S H, Guo L J, et al. Semiconductor-based photocatalytic hydrogen generation. Chemical Reviews, 2010, 110: 6503–6570.
[3] Umebayashi T, Yamaki T, Itoh H, et al. Analysis of electronic structures of 3d transition metal-doped TiO2 based on band calculations. Journal of Physics and Chemistry of Solids, 2002,63: 1909–1920.
[4] Nishikawa T, Shinohara Y, Nakajima T, et al. Prospect of activating a photocatalyst by sunlight -A quantum chemical study of isomorphically substituted titania. Chemistry Letters, 1999: 1133–1134.
[5] Yamashita H, Harada M, Misaka J, et al. Degradation of propanol diluted in water under visible light irradiation using metal ion-implanted titanium dioxide photocatalysts. Journal of Photochemistry and Photobiology a-Chemistry, 2002, 148: 257–261.
[6] Chen X, Glans P A, Qiu X, et al. X-ray spectroscopic study of the electronic structure of visible-light responsive N-, C- and S-doped TiO2. Journal of Electron Spectroscopy and Related Phenomena, 2008, 162: 67–73.
[7] Chen X B, Burda C. The electronic origin of the visible-light absorption properties of C-, N-and S-doped TiO2 nanomaterials. Journal of the American Chemical Society, 2008, 130: 5018–5019.
[8] Asahi R, Morikawa T, Ohwaki T, et al. Visible-light photocatalysis in nitrogen-doped titanium oxides. Science, 2001, 293: 269–271.
[9] Umebayashi T, Yamaki T, Itoh H, et al. Band gap narrowing of titanium dioxide by sulfur doping. Applied Physics Letters, 2002, 81: 454–456.
[10] Ho W K, Yu J C, Lee S C. Low-temperature hydrothermal synthesis of S-doped TiO2 with visible light photocatalytic activity. Journal of Solid State Chemistry, 2006, 179: 1171–1176.
[11] Chen X Q, Zhang X W, Su Y L, et al. Preparation of visible-light responsive P F-codoped TiO2 nanotubes. Applied Surface Science, 2008, 254: 6693–6696.
[12] Gai Y Q, Li J B, Li S S, et al. Design of narrow-gap TiO2 for enhanced photoelectrochemical activity: A passivated codoping approach. Physical Review Letters, 2009, 102.
[13] Lei Z B, Ma G J, Liu M Y, et al. Sulfur-substituted and zinc-doped In(OH)3: A new class of catalyst for photocatalytic H2 production from water under visible light illumination. Journal of Catalysis, 2006, 237: 322–329.
[14] Zhang K, Jing D W, Xing C J, et al. Significantly improved photocatalytic hydrogen production activity over Cd1-xZnxS photocatalysts prepared by a novel thermal sulfuration method. International Journal of Hydrogen Energy, 2007, 32: 4685–4691.

[15] Tsuji I, Kato H, Kobayashi H, et al. Photocatalytic H2 evolution reaction from aqueous solutions over band structure-controlled (AgIn)xZn2(1-x)S2 solid solution photocatalysts with visible-light response and their surface nanostructures. Journal of the American Chemical Society, 2004, 126: 13406–13413.
[16] Kudo A, Mikami I. Photocatalytic activities and photophysical properties of Ga2-xInxO3 solid solution. Journal of the Chemical Society-Faraday Transactions, 1998, 94: 2929–2932.
[17] Zou Z G, Ye J H, Arakawa H. Optical and structural properties of the BiTa1-xNbxO4 (0≤x≤1) compounds. Solid State Communications, 2001, 119: 471–475.
[18] Luan J F, Zou Z, Lu M, et al. Structural, optical and photocatalytic properties of new solid photocatalysts, BixIn1-xTaO4 (0<x<1). Materials Chemistry and Physics, 2006, 98: 434–441.
[19] Yi Z G, Ye J H. Band gap tuning of Na1-xLaxTa1-xCoxO3 solid solutions for visible light photocatalysis. Applied Physics Letters, 2007, 91.
[20] Jang J S, Kim H G, Joshi U A, et al. Fabrication of CdS nanowires decorated with TiO2 nanoparticles for photocatalytic hydrogen production under visible light irradiation. International Journal of Hydrogen Energy, 2008, 33: 5975–5980.
[21] Chen Y B, Guo L J. Highly efficient visible-light-driven photocatalytic hydrogen production from water using Cd0.5Zn0.5S/TNTs (titanate nanotubes) nanocomposites without noble metals. Journal of Materials Chemistry, 2012, 22: 7507–7514.
[22] Wang C J, Thompson R L, Baltrus J, et al. Visible light photoreduction of CO2 using CdSe/Pt/TiO2 heterostructured catalysts. Journal of Physical Chemistry Letters, 2010, 1: 48–53.
[23] Fan W Q, Zhang Q H, Wang Y. Boosting supercapacitor performance of carbon fibres using electrochemically reduced graphene oxide additives. Physical Chemistry Chemical Physics, 2013, 15: 2632–2649.
[24] Kim H G, Borse P H, Choi W Y, et al. Photocatalytic nanodiodes for visible-light photocatalysis. Angewandte Chemie-International Edition, 2005, 44: 4585–4589.
[25] Kim H G, Borse P H, Jang J S, et al. Fabrication of CaFe2O4/MgFe2O4 bulk heterojunction for enhanced visible light photocatalysis. Chemical Communications, 2009: 5889–5891.
[26] Wang D F, Zou Z G, Ye J H. Photocatalytic water splitting with the Cr-doped Ba2In2O5/In2O3 composite oxide semiconductors. Chemistry of Materials, 2005, 17: 3255–3261.
[27] Hou J G, Wang Z, Kan W B, et al. Efficient visible-light-driven photocatalytic hydrogen production using CdS@ TaON core-shell composites coupled with graphene oxide nanosheets. Journal of Materials Chemistry, 2012, 22: 7291–7299.
[28] Lewis N S. Chemical control of charge transfer and recombination at semiconductor photoelectrode surfaces. Inorganic Chemistry, 2005, 44: 6900–6911.
[29] Uosaki K, Kita H. Mechanistic study of photoelectrochemical reactions at ap-GaP electrode. Journal of the Electrochemical Society, 1981, 128: 2153–2158.
[30] Khan S U M, Bockris J O. A model for electron transfer at the illuminated p-type semiconductor-solution interface. Journal of Physical Chemistry, 1984, 88: 2504–2515.
[31] Ma W, Han D, Gan S, et al. Rapid and specific sensing of gallic acid with a photoelectrochemical platform based on polyaniline-reduced graphene oxide-TiO2. Chemical Communications, 2013, 49: 7842–7844.
[32] Ma W, Han D, Zhou M, et al. Ultrathin gC3N4/TiO2 composites as photoelectrochemical elements for the real-time evaluation of global antioxidant capacity. Chemical Science, 2014, 5: 3946–3951.

7 Analytical instrument for antioxidant capacity evaluation

After so many years of development, a variety of suitable instrumental analytical techniques combined with the evaluation of their antioxidant capacity has come into existence; for example, optical color analysis-based kits, optical luminescence-based instruments, chromatographic and mass spectrometry instruments based on analytical separation techniques and recently developed photoelectrochemical instruments based on electrochemical potential and amperometric instruments. With these methods, we can detect the composition and content of antioxidants as well as antioxidant capacity of food. (Figure 7.1)

When spectroscopic technique is applied for antioxidant capacity analysis, the detection principle should be the color change before reaction of standard substances and reactive oxygen species. Although optical detection is cheap and simple, it has the disadvantage of excessive time-consumption and color interferences. Chromatographic techniques help in the isolation and quantification of an individual antioxidant; as a result, assessment of concentration of only the individual antioxidant instead of global capacity of foodstuffs is evaluated, which means that this approach limits the synergistic effect among the assay of antioxidant components. The electrochemical methods are of high sensitivity and free of sophisticated pretreatments; however, the working electrode is likely to be fouled, which limits the measuring speed as well as high throughput determination. Therefore, a photoelectrochemical method was introduced to detect the antioxidant capacity, which is rapid, sensitive and pollution-free. In this chapter, we briefly introduce the commonly used analytical instrumentation technology in the field of antioxidant analysis, including optical color kits, luminous analytical instruments, electrochemical instruments and photoelectric analysis instruments.

7.1 Optical kit

A commercial food testing kit (referred to as a food testing kit) is a commercial test system that includes the primary composition of an analytical technology, which determines the target substance in the foodstuff. Compared with the traditional test system, these kits are simple, fast, convenient and suitable for site operation; these kits lower the standards of required operational skills and results can be obtained in a short span of time. As an important criterion of food, detection of antioxidant capacity of foods has become an increased interest in the research area. The kits based on spectroscopy have been developed extensively. A detailed introduction to commercial testing kits of antioxidant capacity is discussed in the following section.

https://doi.org/10.1515/9783110573763-007

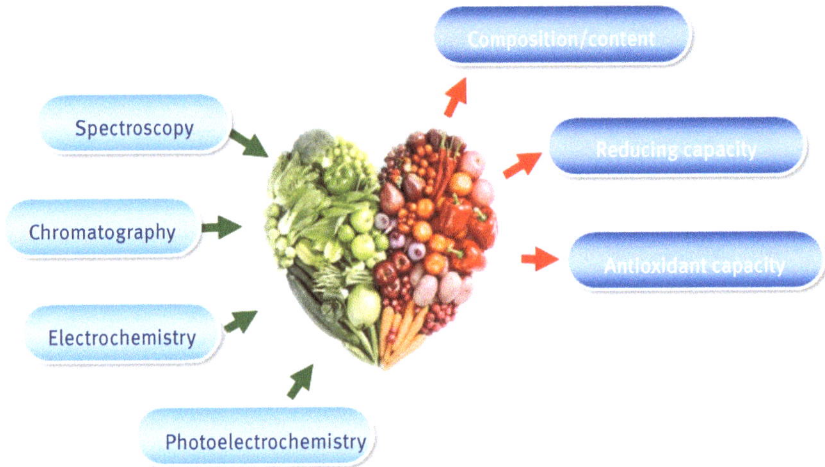

Figure 7.1: Analytical methods of antioxidant assay.

7.1.1 Bathocuproine testing kit

At present, bathocuproine detection kit (Cu^{2+}) has become a universally used method to detect antioxidant capacity. The basic mechanism is shown in Figure 7.2.

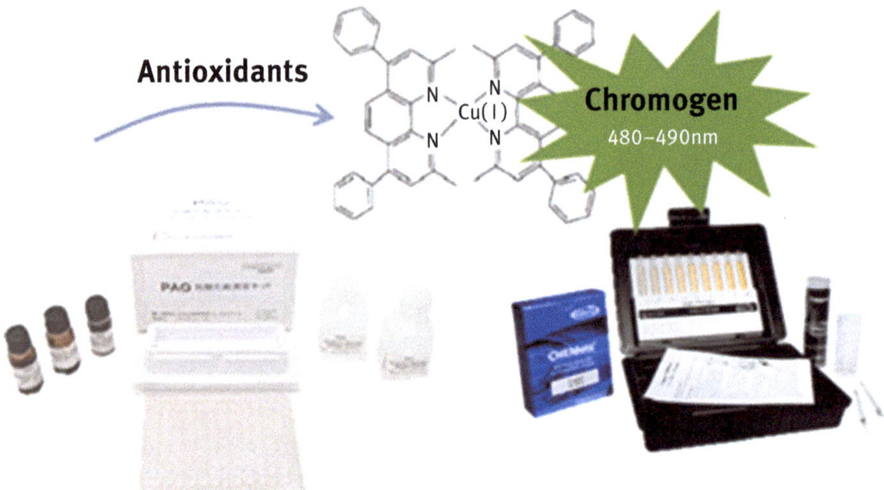

Figure 7.2: The mechanism of Cu^{2+}-based antioxidant capacity test.

Cu^{2+} ions having oxidative properties are supposed to react with antioxidants to produce Cu^{+}. The Cu^{+} can react with the chromogen to produce colored complexes that have maximum absorption at 490 nm. The antioxidant capacity can be determined through the change of absorption.

The OxiSelect MTAC testing kit of Cell Biolabs, Inc is based on the principles mentioned previously, which can be utilized for determining the antioxidant capacity of antioxidants in cell lysates, plasma, serum, urine, tissue homogenates and food extracts. Each kit can provide reagents for determination of 200 blanks, standard samples and actual sample. At the same time, this kit can achieve the assay of both water-soluble and fat-soluble samples.

Using uric acid as a standard substance, the steps are as follows:

(1) Add 20μL of diluted uric acid standard to a 96-well microtiter plate
(2) Use a multichannel pipette to add 180μL of 1x reaction buffer to each well and mix.
(3) The initial absorbance values were measured at 490 nm via the control panel.
(4) Add 50μL of 1x copper ion reagent to each well and incubate for 5 min on an orbital shaker to start the reaction.
(5) Add 50μL of 1x stop solution to each well to terminate the reaction.
(6) Record the absorbance at 490 nm via the control panel.

The results are shown in Figure 7.3.

Uric acid is often used as a standard substance in the bathocuproine testing kit (Cu^{2+}). A suitable linear relationship can be obtained when the uric acid is in a lower concentration range as shown in Figure 7.3 (a). This method is also used to measure the antioxidant capacity of some other antioxidants, such as ascorbic acid, water-soluble vitamins, glutathione, vitamin E, bovine serum albumin (BSA) and butylated hydroxytoluene (BHT), as shown in Figure 7.3 (b). The method can also effectively measure global antioxidant capacity of mixed antioxidants in foodstuff. This reflects the synergistic effect on antioxidants. It can be concluded that an optical kit can be used as a universal method to determine the antioxidant capacity.

7.1.2 ABTS antioxidant assay kit

There are many manufacturers, for example, Zen-Bio, who produce 2,2-*N*-diazo-bis (3-methyl-benzothiazole-6-sulfonyl acid) diammonium salt (ABTS) antioxidant assay kits in the market. The basic mechanism is as follows:

$$\mathrm{HX-FeIII} + \mathrm{H_2O_2} \rightarrow \cdot\mathrm{X-[FeIV=O]} + \mathrm{H_2O}$$

$$\mathrm{ABTS} + \cdot\mathrm{X-[FeIV=O]} \rightarrow \mathrm{ABTS^+} + \mathrm{HX-FeIII}$$

HX-FeIII refers to myoglobin, and X-[FeIV=O] refers to iron (FeIII) myoglobin-free radicals.

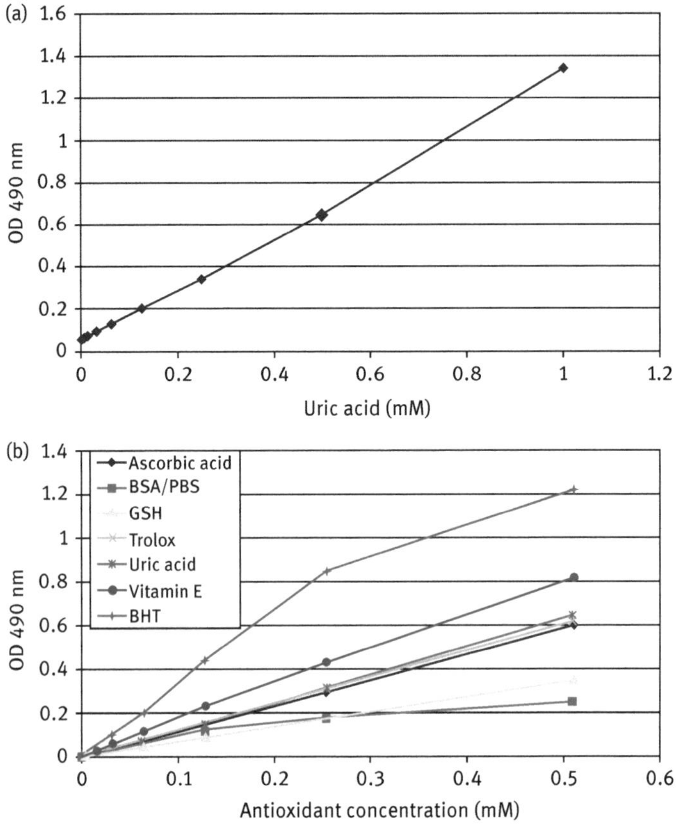

Figure 7.3: (a) Uric acid standard curve and (b) actual antioxidant curve.

The mechanism is based on the fact that myoglobin can react with H_2O_2 to produce iron Fe(III) myoglobin-free radicals, and iron Fe(III) myoglobin-free radicals react with ABTS to generate $ABTS^+$. The $ABTS^+$ solution is green in color, which has maximum absorption at 405 nm. The addition of antioxidants inhibits the formation of $ABTS^+$, which suggests that the antioxidant capacity can be determined on the basis of color change.

Water-soluble vitamin simulator Trolox is used as a standard in ABTS antioxidant assay kit. The detailed procedure is as follows:

(1) Remove the ABTS solution from the refrigerator and allow its temperature to rise to room temperature.
(2) Prepare the Trolox standard solution as shown in Figure 7.4.
(3) Prepare a myoglobin working solution. Remove 25μL of myoglobin storage solution, add 2.475mL buffer solution, shake and place on ice before use.
(4) Add 10μL of actual sample or Trolox standard to each well of the perforated plate, and add 10μL of blank buffer solution as a controlled trial.

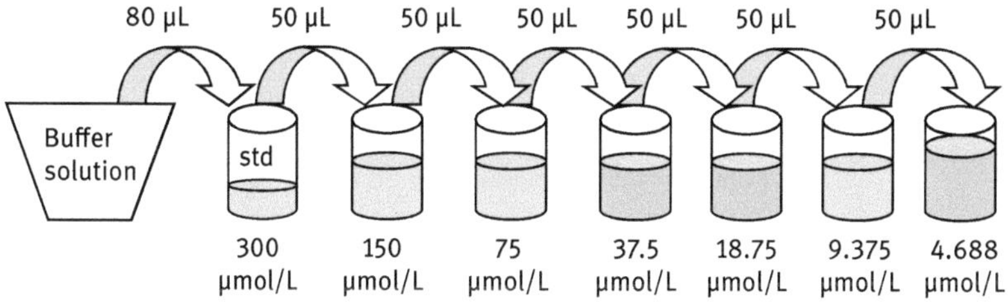

Figure 7.4: Preparation of Trolox standard solution by gradient dilution.

(5) Add 20μL of myoglobin dilution to the wells containing the sample or standard solution.
(6) Add 100μL of ABTS solution to each of the well mentioned above; react at room temperature for 5 min after shaking. Thereafter, 50μL of the stop solution was added to each well.
(7) Record the absorbance at 405 nm via the control panel.

A good linear relationship is obtained using Trolox solution as standard as shown in Figure 7.5(a). This method can also be used to determine the antioxidant capacity of gallic acid, epigallocatechin gallate as well as a mixture of antioxidants in foodstuff, which effectively reflect the synergistic effect.

The two types of assay kits having antioxidant capacities introduced previously are the main kits available on the market, which is simple, convenient and fast. However, the main antioxidants that are mainly used for testing are the small molecular antioxidants, whereas the antioxidant activity of the antioxidant enzyme in human body is of vital significance. The antioxidant capacity of an enzyme is mainly determined by the activity of the enzyme.

7.1.3 Detection methods for biological enzymes

Antioxidant enzymes (glutathione peroxidase, glutathione reductase, peroxidase, superoxide dismutase, etc.) in the organism can effectively eliminate the oxidative substances in vivo through direct or indirect ways to play the role of antioxidants. Therefore, the determination of enzyme activity can effectively reflect the antioxidant capacity in vivo.

As an example, the principle behind the determination of glutathione peroxidase is as follows:

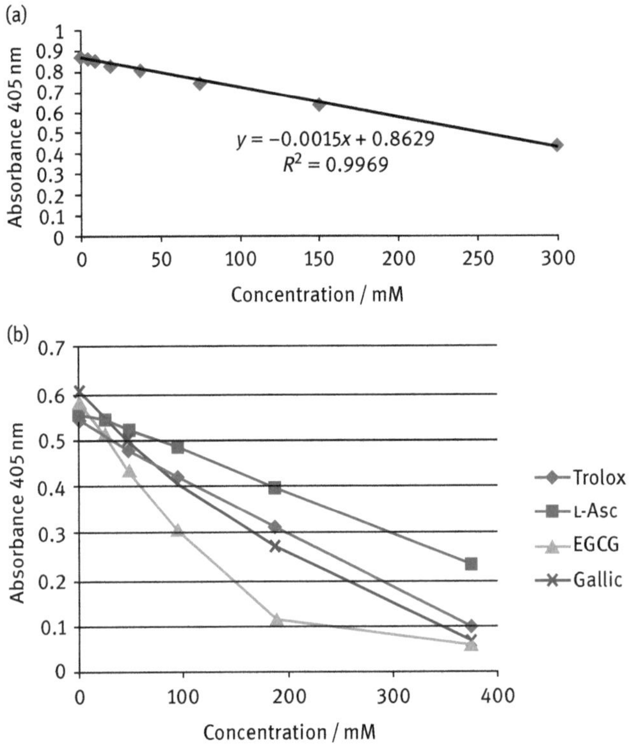

Figure 7.5: a. Trolox standard curve and b. Determination curve of four antioxidants.

$$2GSH \quad + \quad H_2O_2 \xrightarrow{\text{Glutathione Peroxidase}} GSSH \quad + \quad H2O$$

$$GSSH \quad + \beta - NADPH \xrightarrow{\text{Glutathione Reductase}} \beta - NADP \; + \; 2GSH$$

Under the influence of glutathione peroxidase, H_2O_2 can be oxidized and then reduced to glutathione (GSH) to produce oxidized glutathione (GSSG). Subsequently, GSSH can be reduced to GSH by β-NADPH with the help of glutathione reductase. Based on the above-mentioned principle, the activity of glutathione peroxidase can be calculated using the following formula:

$$\text{Enzyme(units/mL)} = \frac{(\Delta A_{340nm}/\min Test - \Delta A_{340nm}/\min Blank)cV_1df)}{\rho V_2}$$

A_{340nm}/minTest is the absorbance of the testing sample at 340nm
A_{340nm}/minBlank is the absorbance of blank sample at 340nm
c is the amount of GSH produced by β-NADPH per μmol
V_1 is the total volume (mL) of the testing system

df is the dilution factor
ρ is 6.22 (the millimeter extinction coefficient of β-NADPH at 340 nm)
V_2 is the volume (mL) of enzyme.

Similarly, the enzymatic activity of other antioxidant-related biological enzymes can also be obtained by using a similar method. The use of optical kits for the detection of antioxidant capacity of foodstuffs is simple, fast and convenient. It is suitable for field operation and gives result in a very short span of time. However, the kit method is mostly qualitative and semiquantitative, whereas the method based on instruments is useful for obtaining more accurate results. Currently, besides the traditional LC/LC-MS which can be used for the separation and analysis of antioxidants, there are three main analytical instruments for detection of antioxidant capacity: (1) luminous method; (2) potential method; photoelectrochemical method.

7.2 Antioxidant potential of a luminous detector

The detectors for the determination of antioxidant capacity with luminous method are mainly produced by Analytik Jena, Germany and Tohoku Electronics Co., Ltd, Japan.

7.2.1 Jena Germany (http://www.analytik-jena.de/)

The luminescence method for determining the antioxidant capacity was proposed by Igor et al. Based on this method, they developed an instrument called Analytikjena PHOTOCHEM® for measuring antioxidant capacity (Figure 7.6).

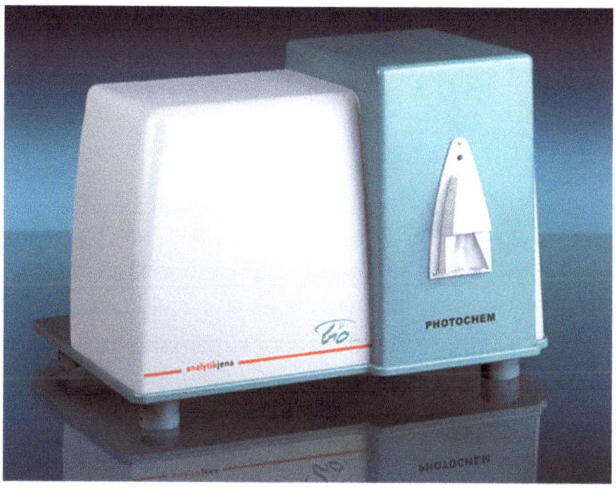

Figure 7.6: Analytikjena PHOTOCHEM® antioxidation activity detector, Jena, Germany.

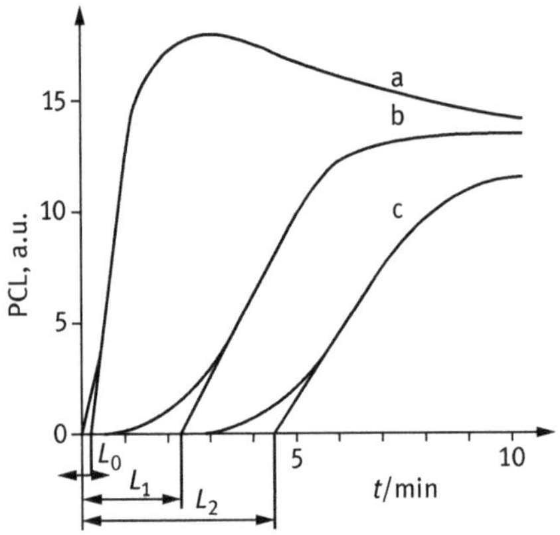

Figure 7.7: Chemiluminescence spectroscopy results. a, b and c are the assay system with 0, 2 and 4μL plasma samples (L_0, L_1, L_2 refer to the delay time).

(1) Mechanism: The principle of chemiluminescence (PCL) is mainly based on the chemical reaction with UV-A (365 nm) as an excitation light, including two reaction steps:

$$S + hv \rightarrow S^* \tag{7.1}$$

$$S^* + O_2 \rightarrow S^+ + O_2^- \tag{7.2}$$

$$S^* + O_2 \rightarrow S + 1O_2 \tag{7.3}$$

S refers to the photon, hv refers to the ultraviolet and S* refers to the excited photon.

The company designed different light-emitting paths based on different photons. Luminol (L) will be described as a photosensitive molecule.

$$LH^- + hv_1 \rightarrow L^{-\cdot} + O_2^{-\cdot} \rightarrow N_2 + AP^{2-} + hv_2 \tag{7.4}$$

$$L^{-\cdot} + AOH \rightarrow AO^{\cdot} + LH^- \tag{7.5}$$

$$O_2^{-\cdot} + AOH \rightarrow AO^{\cdot} + HO_2^- \tag{7.6}$$

hv_1 refers to the excitation light, hv_2 refers to the emitted light, AP^{2-} refers to the aminophthalic anion, AOH refers to the antioxidant and AO refers to the oxidative antioxidant.

The antioxidants can react with luminol-free radicals or superoxide radicals to reduce the luminescence intensity. Therefore, the antioxidant activity can be determined depending on the change in luminescence intensity.

The antioxidant capacity can be determined by two ways: (i) delay time and (ii) integration area.

Make a tangent to the chemiluminescence diagram, and the focus of the tangent and the time axis is the delay time. The results is $L=L_1-L_0$ (L_1 refers to the delay time of the samples; L_2 refers to the delay time of the blank samples). Different delay times can be obtained from samples of different concentration. We can prepare a standard curve of the relationship between the concentration and the delay time, and with this the concentration of actual samples can be determined.

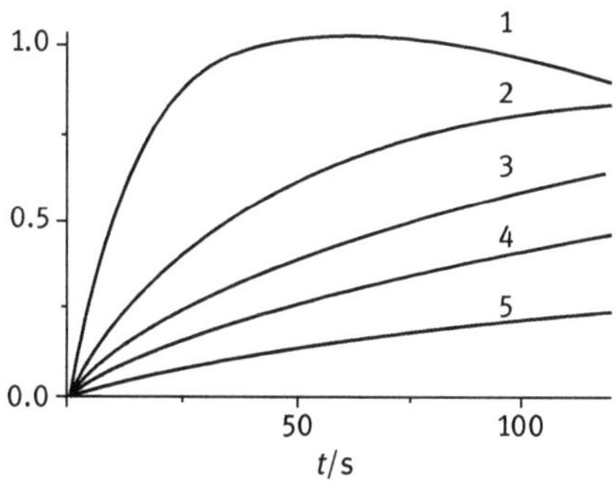

Figure 7.8: Chemiluminescence spectrum (1→5) with 0, 20, 40, 60 and 80 μL plasma extract.

Another method for determination is through integral area, which is calculated by the area of the luminescence curve of the first 2 minutes (Figure 7.8). This is expressed by the luminescence suppression parameter.

$$I = 1 - S_1/S_0$$

S_1 refers to the integral area of the samples of the first 2 minutes; S_0 refers to the integral area of the blank samples of the first 2 minutes. The antioxidant capacity can be determined by suppression parameters.

PHOTOCHEM® photochemical luminescence analyzer can determine the global antioxidant capacity of mixtures as well as individual antioxidant or superoxide dismutase. Therefore, this instrument can be applied to research and routine analysis in various fields, such as food technology, chemical engineering, agricultural analysis, pharmaceuticals and biochemistry.

1 The main technical parameters of instrument system

The main structure of the instrument is shown in Figure 7.9. The buffer and sample are driven to the reaction cell by the peristaltic pump. The signal after reaction is collected by the monitor and then stored after transformation is complete.

2 Technical characteristics of instrument system

(a) Simultaneous determination of water-soluble and oil-soluble antioxidants
(b) Detection limit reaches nmol/L level
(c) Short measuring time

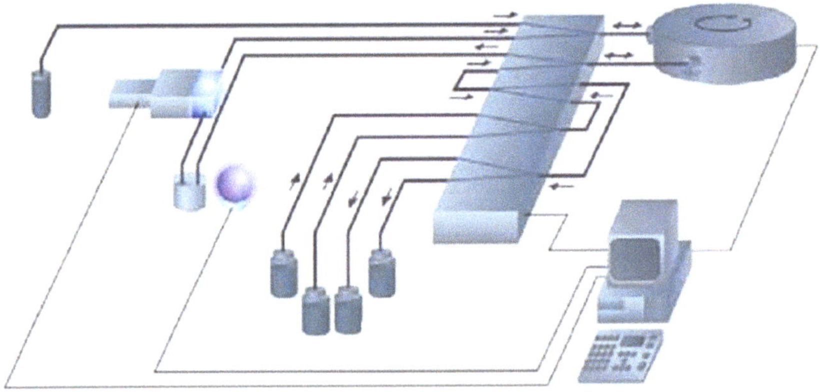

Figure 7.9: Diagram of PHOTOCHEM® antioxidant detector.

(d) Synergistic effect of antioxidant components can be reflected
(e) The ability to detect the antioxidant enzyme activity of enzymes, such as SOD

However, the instrumentation system is more expensive, whereas the measurement is nonselective and requires complex pretreatment processes. At the same time, the instrument is supposed to work under alkaline conditions, which deviates from the physiological environment and the color of the luminescent assay have great interference on the final results.

7.2.2 Tohoku Electronic Industrial (http://www.tei-c.com)

Northeast electronics industry developed optical detection equipment to measure the antioxidant capacity based on obvious luminous mechanism (Figure 7.10).

1 Mechanism of the method

This system uses thermal excitation instead of UV excitation; the principle is as follows:

$$R-N=N-R \xrightarrow{37^{\circ}C} 2R^{\bullet} + N_2\uparrow$$

$$R^{\bullet} + O_2 \rightarrow ROO^{\bullet}$$

$$ROO^{\bullet} + LH^{\bullet} \nearrow ROOH + L^{-\bullet}$$

$$ROO^{\bullet} + LH^{\bullet} \searrow R^{\bullet} + LHOO^{-} \searrow N_2\uparrow + AP^{*} \searrow AP^{2-} + h\nu$$

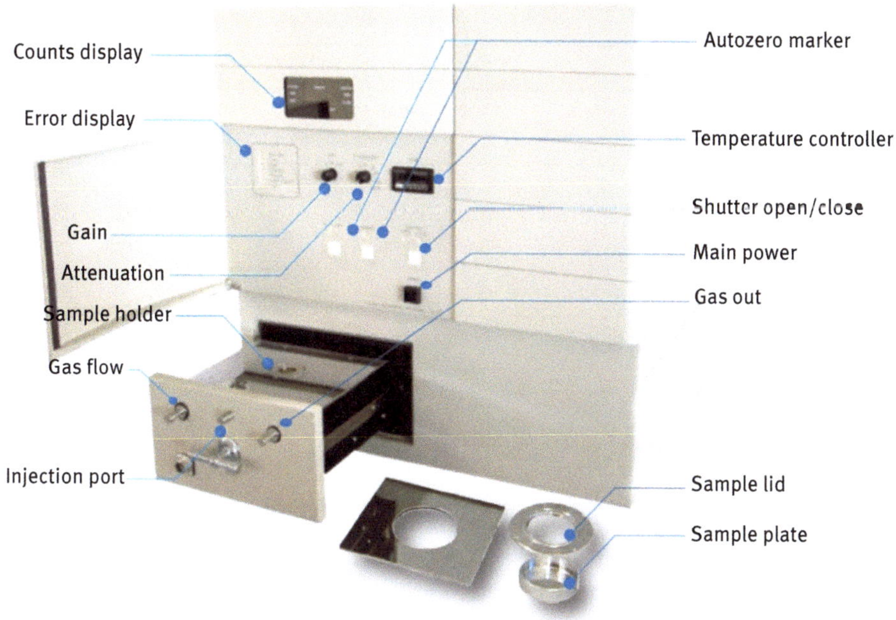

Figure 7.10: The antioxidant measuring instrument produced by Japan Northeast electronics industry.

R^-—N=N—R is an azide compound
LH$^\cdot$is luminol
AP^{2-} is an aminophthalic acid anion
hν is the chemical excitation

The most often choice of azide compounds can be 2,2′-azobis (2-amidinopropane) dihydrochloride, azobisisobutyronitrile and azobisisobutyronitrile. Azide compounds can decompose with the release of nitrogen-containing radicals at a certain temperature ($R^\cdot$), and then R reacts with oxygen to form oxygen-containing radicals ($ROO^\cdot$). $ROO^\cdot$ can oxidize luminol to form aminophthalic acid activator (AP*), which continues to decompose with the generation of aminophthalic anions, while producing chemiluminescence. Because of the strong reducing property of antioxidants, they prefer to react with free radicals in the above-mentioned system to decrease the light intensity or increase the light emission time. Based on this mechanism, the antioxidant capacity can be measured. Simple instrument systems for thermal excitation can be designed because of the lack of excitation light. The main technical parameters of the instrument system are listed in Table 7.1.

Table 7.1: Performance indicators of different models of antioxidant capacity detectors.

Model	CLA-FS3	CLA-ID3	CLA-FL2
Detecting wavelength	300~850 nm, Middle wavelength: 420nm	300~650nm, Maximum quantum yield: 430nm	300~650nm, Maximum quantum yield: 430nm
Mode of measurement	Luminous intensity (counts/s) and spectrum mode	Luminous intensity (count/s) without Spectrum mode	Luminous intensity (count/s) without Spectrum mode
Light filter	20	0	0
Temperature	Room temperature ~160 °C	Room temperature ~100 °C	Room temperature ~50 °C
Sample chamber	CLS-ST3, MX3, FL, LA1	CLS- MX3, FL, LA1	CLS-FL
Port	RS-232	RS-232	None
Size/quality	520(W)×420(D)×547(H)mm ~50kg	310(W)×420(D)×524(H)mm ~35kg	310(W)×420(D)×524(H)mm ~35kg

2 Technical characteristics of various instrument systems:

Apart from measuring the antioxidant capacity of the antioxidant, these instruments also measure three additional classes of substances:

(a) Determination of the early oxidation state
(b) Measuring oxidative substances
(c) Ultrasensitive fluorescence determination

3 Operating steps for instrument system

For example, the antioxidant capacity measurement of beverages:

Configure ABAP, luminol and phosphate buffer solutions of certain concentrate

(a) Mix the three solutions with certain percentage and keep at 37 °C for 15 min
(b) Add different concentrations of Trolox standard solution
(c) Chemiluminescence was measured using the CLA-FS3 detector
(d) Calculate the standard curve
(e) Add a certain volume of beverages (grape juice, green tea) to replace Trolox standard solution in step (c)
(f) Repeat step (d) to calculate the antioxidant capacity of the beverage

The results are shown in Figure 7.11

As shown in Figure 7.11, free radicals oxidize light-emitting agents to form luminous agent-free radicals while producing chemiluminescence, which can be recorded by the detector. After the addition of the antioxidant, some free radicals are eliminated and the luminescence is inhibited. When the antioxidant is completely

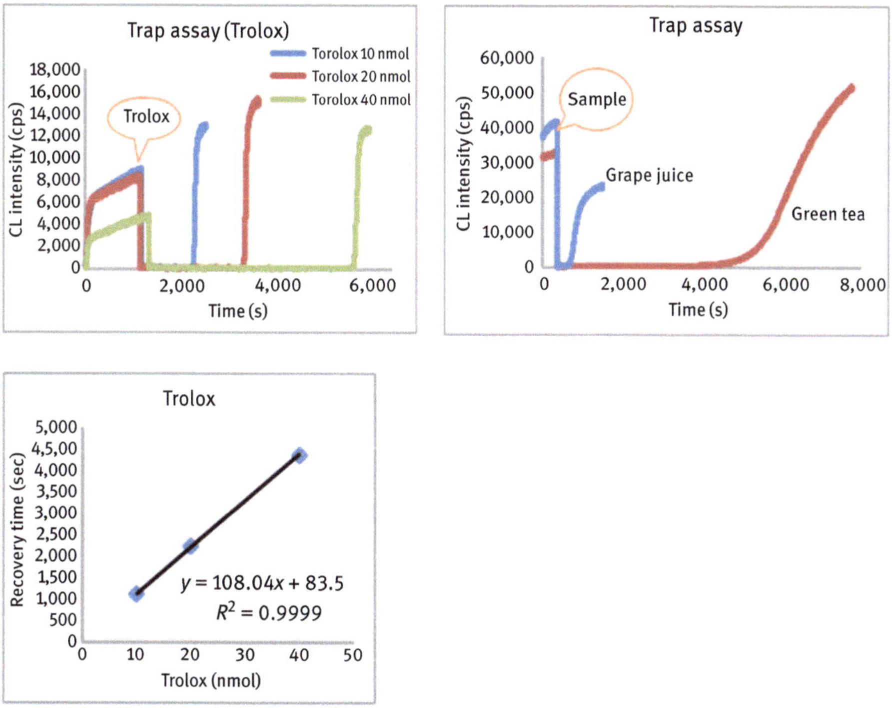

	Recovery time(s)
Grape juice	473
Green tea	5,932

$Y = 108.04x + 83.5$

Grape juice : 3.6 nmol
Grape tea : 54.1 nmol

Figure 7.11: Determination of antioxidant capacity of Trolox, grape juice and green tea via thermal excitation method.

consumed, chemiluminescence reappears, and thus the time without chemiluminescense is called the induction period. The induction period is usually utilized for quantifying the antioxidant capacity, and the experimental results are expressed as Trolox equivalents.

As concluded from Figure 7.11, the induction time of chemiluminescence also increases with a linear relationship as the concentration of Trolox increases. Compared with the same volume of green tea and grape juice, we observe that green tea showed antioxidant capacity 15 times as high as grape juice.

Chemiluminescence initially utilized the luminol system, but the luminous intensity is weak, so the luminol is gradually replaced by luciferin and bioluminescent protein Pholasin, which can provide continuous, stable and strong luminescence signal.

The chemiluminescence method can measure the antioxidant capacity of antioxidants in samples of large quantities. However, the sample requires complicated pretreatment process. Meanwhile, the colored substance that does not react completely can easily absorb the light emitted by the chemical reaction and affect the final result.

7.3 Potential method

Russian scientist K. Z. Brainina and group developed a potential detector for measuring antioxidant capacity based on their research [8–10] (Figure 7.12).

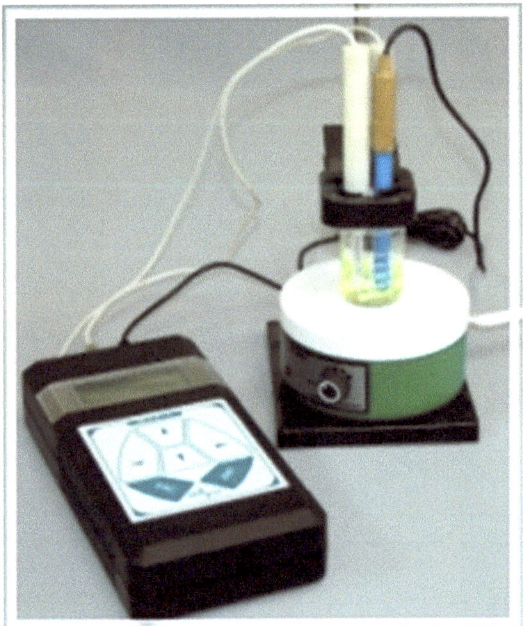

Figure 7.12: MPA-1 multichannel antioxidant capacity potentiometer.

1 Mechanism of the system

The mechanism is based on the principle that addition of antioxidants can change the concentration of the media, and thus can change the redox potential of the system. Therefore, the antioxidant capacity can be measured by a change in the redox potential of the system.

$$E = E_0 + b\lg C_{ox}/C_{red}$$

$$E_1 = E_0 + b\lg(C_{ox} - X)/(C_{red} - X)$$

$$AOA = \pm(C_{ox} - \alpha C_{red})/(1 + \alpha)$$

E is the potential of the system before the addition of antioxidants; E_0 is the standard redox potential of K_4 [Fe (CN_6)]/K_3 [Fe (CN_6)]; C_{ox} is the concentration of K_3 [Fe (CN_6)]; C_{re} is the concentration of K_4 [Fe (CN_6)];E_1 is the potential of the system after addition of antioxidants; X is the concentration of antioxidants; AOA is antioxidant capacity; α is $10^{(E1\text{-}E)/b}$, b=2.3RT/nF.

2 Main technical parameters of the instrument system

(a) Potential measurement range of +4.0V~−4.0V
(b) The detection limit is 10^{-6}mol/L
(c) A single sample measuring time is 3~5min
(d) Determine the other 26 kinds of ions simultaneously (Cl^-, F^-, NO_3^-, SO_4^{2-}, CN^-, Br^-, I^-, Ca^{2+}, Mg^{2+}, K^+, Na^+, S^{2-}, NO_2^-, NH_4^+, Mg^{2+}, HPO_4^{2-}, Ba^{2+}, CNS^-, Fe^{3+}, ClO_4−, CO_3^{2-}, Ag^+, Cd^{2+}, Cr^{6+}, Cu^{2+}, Hg^{2+})
(e) The power system is 220V AC or battery
(f) Size (mm): 215 (l) × 100 (w) × 40 (h)
(g) Quality: 0.4kg
(h) The interface is liquid crystal (4-line, 16-alphanumeric characters and 6-key keyboard)
(i) Customers can choose the software mode that is automatically activated

3 Technical characteristics of the instrument system

(a) A single sample measuring time is 3~5min
(b) Noncomplicated operation steps
(c) Low cost
(d) Operating software language is English

4 Operating steps

(a) Configure K_3 [Fe (CN_6)] and K_4 [Fe (CN_6)] solution of certain concentration, and then mix them with certain percentage
(b) Measure the potential of the above-mentioned mixture of solutions
(c) Add different concentrations of standard solution to the mixture of solution, and record the corresponding potential
(d) Make the standard curve based on the concentration and potential
(e) Add actual sample solution to the mixed solution prepared in step (1)
(f) The antioxidant capacity of actual sample is measured based on a standard curve.

Using the potential method to determine the antioxidant capacity is simple and inexpensive, but the greatest challenge for this instrument is the pollutants and absorption of the working electrode.

7.4 Antioxidant activity of a photoelectrochemical detector

Antioxidant capacity of the photoelectrochemical detection method was first reported by the Changchun Institute of Applied Chemistry, Chinese Academy of Sciences [11–13]. As introduced in Chapter 6, the mechanism of this method is as follows: semiconductor materials generate electrons and holes under appropriate light. Under a certain applied potential, electrons can be transferred to the working electrode to generate photocurrent. Holes generated by the light oxidize antioxidants to receive electrons; electrons are then excited by light and then amplified to generate photocurrent. The antioxidant capacity can be measured on the basis of intensity of the photocurrent.

Based on the photoelectrochemical detection analysis, researchers have developed related instruments. The researchers first used utg-C_3N_4/TiO_2 as a photoelectric material to design static photoelectrochemical detection cell (Figure 7.13).

Various antioxidants have shown good response with the use of static photoelectrochemical detection cell; however, the sensitivity decreases in continuous measurement because of adsorption of the product (Figure 7.14 (a)). To overcome this problem, the group further designed a photoelectrochemical flow cell (Figure 7.15). The photocurrent in continuous measurement was kept constant as expected, which caused the continuous work of the semiconductor-modified electrode and increased reproducibility (Figure 7.14 (b)). In addition, the application of flow cells increases the substance transfer capability as well as the photocurrent response at the same concentration, and thus increases the sensitivity of the measurement [14].

Based on the above-mentioned research, the group integrated the light source, flow cell, power and detector to develop a new photoelectrochemical antioxidant capacity analyzer (Figure 7.16).

The main technical parameters of the system:

1 Main technical characteristics

(a) Take visible LED light as the excitation source
(b) The use of external excitation light can effectively reduce the interference of colored substance
(c) The use of flow injection mode not only reduces the poisoning of the electrode, but also reduces the sample amount
(d) The optimized two-electrode control acquisition system can effectively overcome some shortcomings, such as the leakage of reference electrode, potential drift and difficulty to integrate
(e) Short operating time and sensitive response
(f) Nonpolluting injection buffer solution – PBS solution

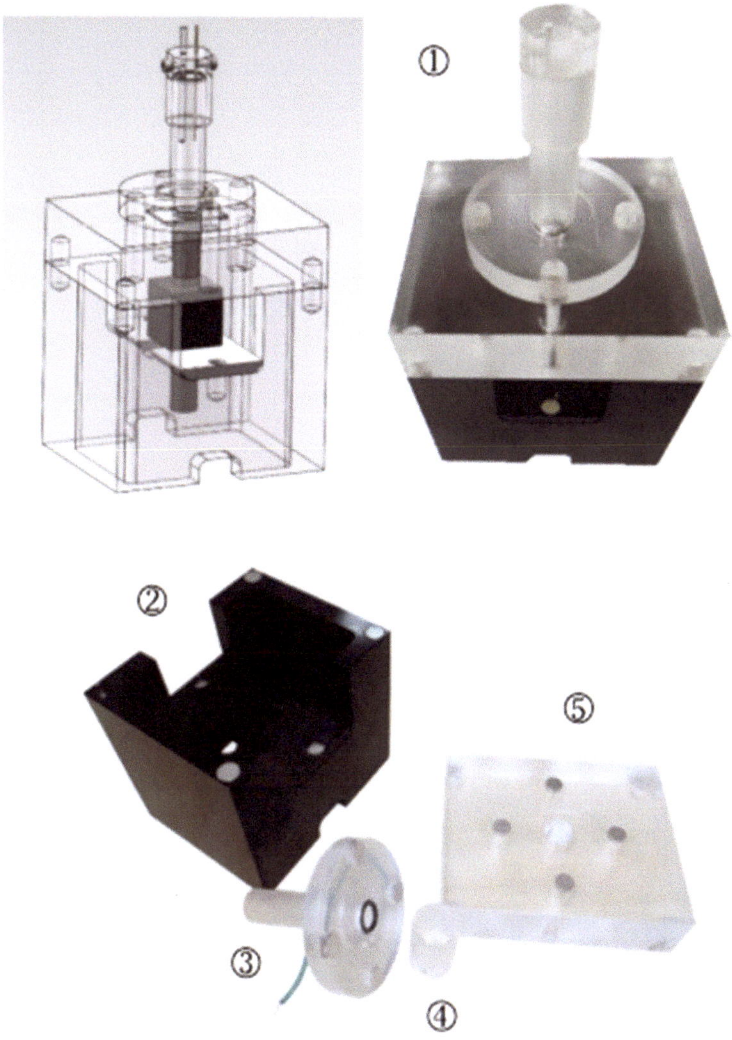

Figure 7.13: ① Static photoelectrochemical detection cell structure and a full set of devices; static photocell consists of ② devise base, ③ static pool, ④ pool cover and ⑤ pool base.

2 Operating steps of instrument system

(a) Configure PBS buffer solution and gallic acid standard solution
(b) Continuous PBS injection mode was used to balance the baseline
(c) Determine the photoelectrochemical signal of different concentration of gallic acid, and prepare the standard curve
(d) Wash the instrument with PBS to prepare for the next assay

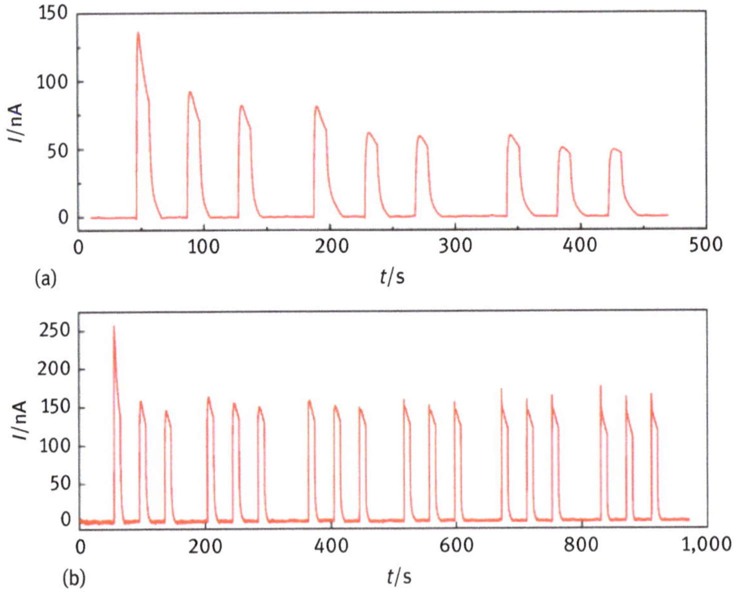

Figure 7.14: Continuous response of photocurrent on utg-C_3N_4/TiO_2 modified electrode. (a) static cell; (b) flow cell.

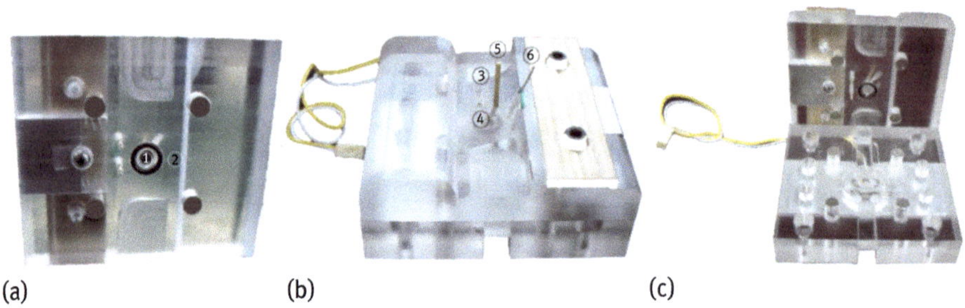

Figure 7.15: Photoelectrochemical flow cell. (a) Internal structure; (b) external structure; (c) integral structure. (① the couple electrode; ② the working electrode; ③/④ the inlet/outlet; ⑤⑥ the lead.).

We use the newly developed instrument to determine the four antioxidants: (1) gallic acid, epicatechin gallate, theophylline and caffeic acid in a two-electrode system. Respectively, they showed a good linear relationship in the range of 11.10~343.35 µmol/L, 5.55~54.94 µmol/L, 8.88~425.53 µmol/L and 5.55~108.70 µmol /L. The method also has a good response to quercetin, to xanthin, rutin, water-soluble vitamin E analogs, ascorbic acid and glutathione (Figure 7.17). The instruments can also

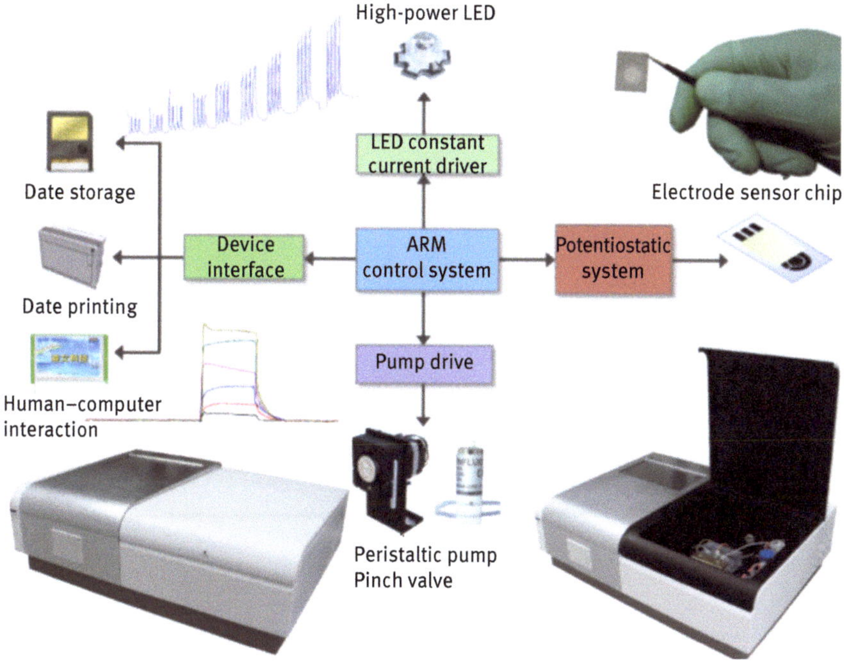

Figure 7.16: Photoelectrochemical antioxidant activity flow detector and system structure.

be used to measure the antioxidant capacity of antioxidants in tea, red wine, coffee and various beverages.

The instrument was used to evaluate the global antioxidant activity of 22 beverages and 16 common fruits (Table 7.2). Because most fruits contain a large amount of antioxidants, beverages containing fruits show strong antioxidant capacity. Blueberries and grapes are widely recognized as fruits containing antioxidants, so the beverages prepared from these two fruits exhibit strong antioxidant capacity, whereas vitamins with water contain more polyphenols, showing super antioxidant capacity. Other beverages have a relatively low antioxidant capacity, especially carbonated beverages exhibit low antioxidant capacity.

Compared with optical method, electrochemistry, chromatography and other analytical instruments, photoelectrochemistry overcome some great limitations. Compared with luminescence analysis and optical kits, this method overcomes the interference caused by colored substances and the long testing time. Because the instrument uses disposable chip so the electrodes are not easily fouled, thus increasing the reproducibility. Compared with liquid chromatography and other separation analysis technology, this method can perfectly reflect the synergy among the antioxidants.

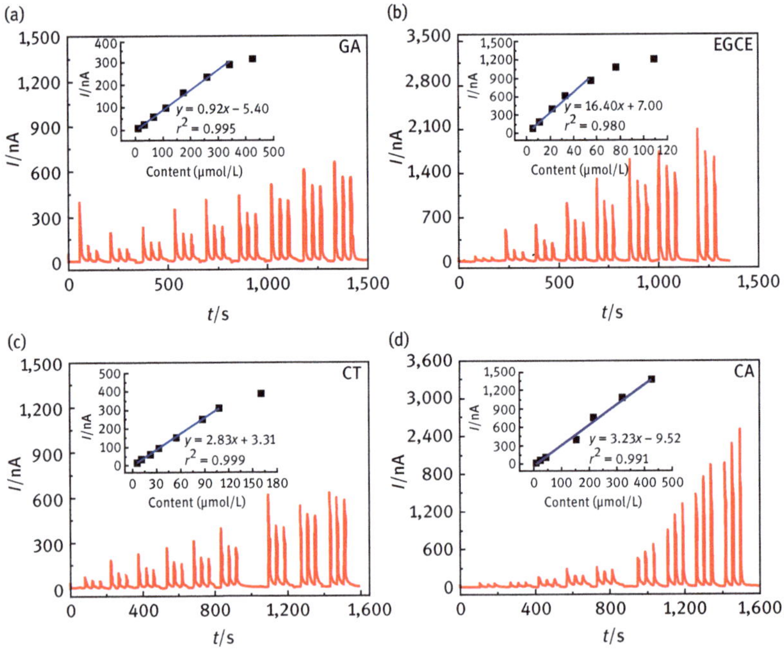

Figure 7.17: The relationship between photocurrent and antioxidant capacity of gallic acid (a), epicatechin gallate (b), theophylline (c) and caffeic acid (d). The inset (a), (b), (c) and (d) are the linear equations of GA, EGCG, CT and CA.

Table 7.2: Antioxidant capacity photoelectric test results of common fruit and juice.

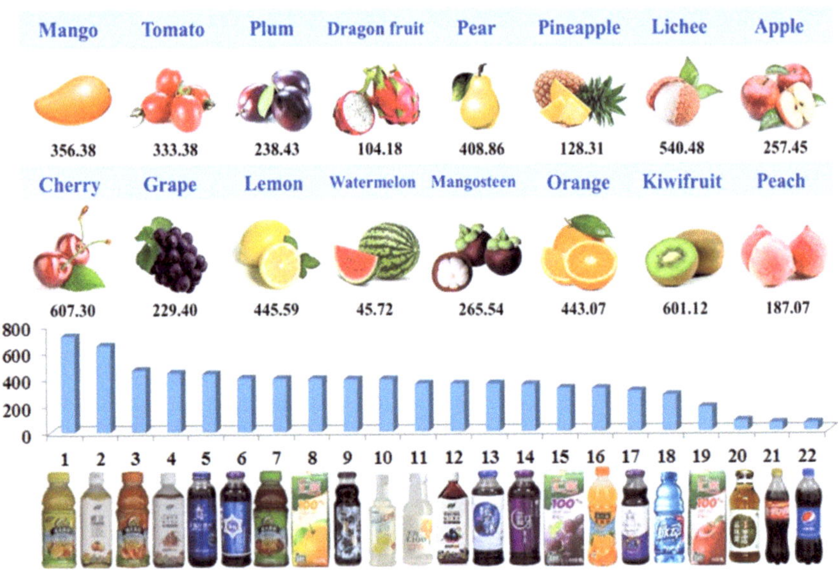

Mango	Tomato	Plum	Dragon fruit	Pear	Pineapple	Lichee	Apple
356.38	333.38	238.43	104.18	408.86	128.31	540.48	257.45
Cherry	**Grape**	**Lemon**	**Watermelon**	**Mangosteen**	**Orange**	**Kiwifruit**	**Peach**
607.30	229.40	445.59	45.72	265.54	443.07	601.12	187.07

References

[1] Lewin G, Popov I. Photocheluminescent detection of antiradical activity; III: A simple assay of ascorbate in blood plasma. Journal of Biochemical and Biophysical Methods, 1994, 28: 277–282.

[2] Popov I, Lewin G. Photochemiluminescent detection of antiradical activity. VI. Antioxidant characteristics of human blood plasma, low density lipoprotein, serum albumin and aminoacids during in vitro oxidation. Luminescence, 1999, 14: 169–174.

[3] Popov I, Lewin G. Photochemiluminescent detection of antiradical activity. VII. Comparison with a modified method of thermo-initiated free radical generation with chemiluminescent detection. Luminescence, 2005, 20: 321–325.

[4] Popov I, Volker H, Lewin G. Photochemiluminescent detection of antiradical activity. V. Application in combination with the hydrogen peroxide-initiated chemiluminescence of blood plasma proteins to evaluate antioxidant homeostasis in humans. Redox Report, 2001, 6: 43–48.

[5] Popov I N, Lewin G. Photochemiluminescent detection of antiradical activity: II. Testing of nonenzymic water-soluble antioxidants, Free Radical Biology and Medicine. Free Radical Biology and Medicine, 1994, 17: 267–271.

[6] Popov I N, Lewin G. Photochemiluminescent detection of antiradical activity; IV: testing of lipid-soluble antioxidants. Journal of Biochemical and Biophysical Methods, 1996, 31: 1–8.

[7] Popov I N, Lewin G, Vonbaehr R. Photocheluminescent detection of antiradical activity; III: a simple assay of ascorbate in blood plasma. Biomedica Biochimica Acta, 1987, 46: 775–779.

[8] Ivanova A V, Gerasimova E L, Brainina K Z. Potentiometric study of antioxidant activity: Development and prospects. Critical Reviews in Analytical Chemistry, 2015, 45: 311–322.

[9] Brainina K Z, Ivanova A V, Sharafutdinova E N, et al. Potentiometry as a method of antioxidant activity investigation. Talanta, 2007, 71: 13–18.

[10] Shpigun L K, Arharova M A, Brainina K Z, et al. Flow injection potentiometric determination of total antioxidant activity of plant extracts. Analytica Chimica Acta, 2006, 573: 419–426.

[11] Ma W, Han D, Gan S, et al. Rapid and specific sensing of gallic acid with a photoelectrochemical platform based on polyaniline-reduced graphene oxide-TiO2. Chemical Communications, 2013, 49: 7842–7844.

[12] Ma W, Han D, Zhou M, et al. Ultrathin gC3N4/TiO2 composites as photoelectrochemical elements for the real-time evaluation of global antioxidant capacity. Chemical Science, 2014, 5: 3946–3951.

[13] Wang L, Ma W, Gan S, et al. Engineered photoelectrochemical platform for rational global antioxidant capacity evaluation based on ultrasensitive sulfonated graphene-TiO2 nanohybrid. Analytical Chemistry, 2014, 86: 10171–10178.

[14] Montgomery E R, Murray R W. Trace analysis of neutral organic species using thin-layer electrochemistry. Analytica Chemical Acta, 1996, 321: 195–199.

Outlook

The evaluation method of antioxidant capacity varies because of the mechanism of detection and conditions of reactive oxygen species reactions of the probe molecule. The selection of the specific detection method should be based on factors such as the testing conditions and the substrate. What is more important is that the concentration of antioxidants should be as close as possible to the actual level in the biological system to evaluate the practical test results. It is because the test methods are diverse and that there is no comprehensive evaluation for the antioxidant capacity at present; thus, we cannot consider antioxidant capacity as a nutritional indicator of food. Therefore, research on the standardization of food antioxidant test methods is an important technical supplement in the field of evaluation of food nutrition. First, the standardized test method can be used to guide the practical application of other test methods. Second, it can make a meaningful comparison of the antioxidant levels of food or commercial samples. Third, it can effectively monitor food quality standards and human health.

In view of the theory and application value of the above standardized test methods, we should focus on the following factors in the study of specific methods: (1) potential applications in the actual chemical reaction; (2) the source of free radicals in the biological body; (3) detection methods have a clear chemical reaction mechanism; (4) effective identification of interferences; (5) easy to operate; (6) convenient and applicable; (7) high reproducibility; (8)applicable to a variety of sources of free radicals and water-soluble or oil-soluble antioxidants; (9) suitable for high-throughput routine analysis detection.

In addition, for complex systems containing a variety of antioxidant components, the global antioxidant capacity is derived from the combined effect of each component. The common natural antioxidants (mainly in fruits and vegetables) are flavonoids, vitamins, indoles polyphenols, and so on. The commonly used artificial antioxidants mainly include tert-butyl hydroxyanisole (BHA), 2,6-di-tert-butyl-p-cresol (BHT), tert-butylhydroquinone (TBHQ) and propyl gallate. For the thirst of food safety combined with animal tests found that synthetic antioxidants may have potential cumulative toxicity and people are increasingly interested in natural "health food". How many "natural ingredients" are there in processed foods or cosmetics? This requires measurement of antioxidant capacity of different components in the system on the basis of understanding of the total antioxidant capacity. Then draw the results into the corresponding map, that is, antioxidation fingerprint analysis. Taking red wine as an example, the total conventional antioxidant capacity analysis can assess the overall antioxidant capacity of the system; however, it fails to figure out the percentage of natural or artificial antioxidants. The antioxidation fingerprint analysis based on artificial neural networks and chemometrics will effectively solve this problem (Figure 8.1). Compared to the comprehensiveness and the integrity of current total antioxidant capacity analysis, antioxidant fingerprinting technology features hierarchy and relevance, which helps to

https://doi.org/10.1515/9783110573763-008

obtain composition and content of food antioxidant system as comprehensive as possible. The quantitative relationship based on this system will be of great significance in the fields of food quality control, nutrition, screening of effective ingredients and food identification.

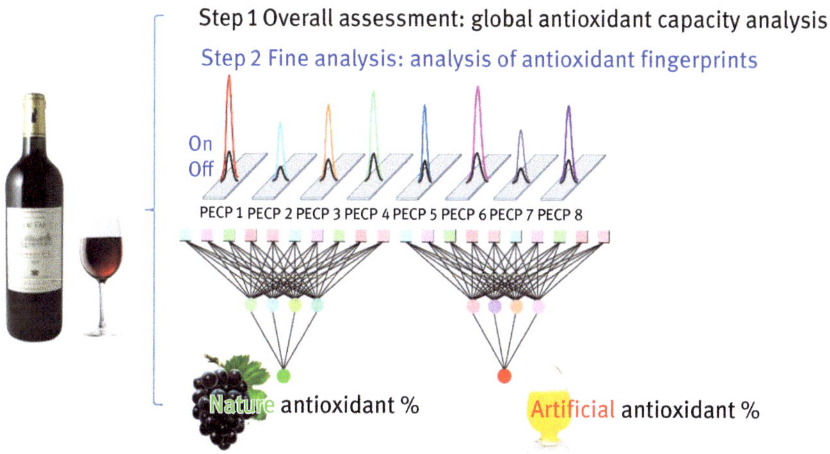

Figure 8.1: Development trend of antioxidant capacity evaluation Based on Photoelectrochemical technique.

The antioxidants in the diet helps to clear excess reactive oxygen species in the human body, so they have importance in the research on relationship between the antioxidants and the oxidative stress levels. The various antioxidant capacity assays described above are important guidelines for clinical testing as long as we combine the biocompatibility of the assay with an effective oxidative stress marking substance. In addition to considering the effectiveness of the detection method, special attention should be paid to the influence of interfering components in the actual biological samples such as plasma, human tissue fluid or urine. Therefore, effective pretreatment process of samples is a crucial step. The synergistic effect of the testing method also provides an effective way to evaluate the ability of antioxidants to reduce oxidative stress levels; meanwhile, it has an important guiding significance for the prevention of human diseases.

Bei Fragen zur Produktsicherheit wenden Sie sich bitte an:
If you have any questions regarding product safety,
please contact:

Walter de Gruyter GmbH
Genthiner Straße 13
10785 Berlin
productsafety@degruyterbrill.com